Practical dam analysis

Max A. M. Herzog
Consulting Engineer

Thomas Telford

Published by Thomas Telford Publishing, Thomas Telford Limited, 1 Heron Quay, London E14 4JD.
URL: http://www.t-telford.co.uk

Distributors for Thomas Telford books are
USA: ASCE Press, 1801 Alexander Bell Drive, Reston, VA 20191-4400
Japan: Maruzen Co. Ltd, Book Department, 3–10 Nihonbashi 2-chome, Chuo-ku, Tokyo 103
Australia: DA Books and Journals, 648 Whitehorse Road, Mitcham 3132, Victoria

Herzog, Max: Elementare Talsperrenstatik mit 26 Zahlenbeispielen von ausgeführten Talsperren im Vergleich zum gemessenen Verhalten.
1. edition 1998
© Werner Verlag GmbH & Co KG, Düsseldorf
ISBN 3-8041-2070-9

This edition 1999

A catalogue record for this book is available from the British Library

ISBN: 0 7277 2725 7

Typeset by Santype International, Salisbury.
Printed and bound in Great Britain by MPG Books Limited, Bodmin.

Preface

With the help of computers and finite element analysis, practically all types and shapes of dam can now be assessed, even including studies of crack formation and seepage. Nevertheless, there is an acute demand for *simple* approximate methods of analysis for design studies, preliminary designs, estimates of quantities, checking and teaching.

In this book, those approximate methods of analysis which have proved useful in the author's more than 40 years of practice are presented and compared with *in situ* measurements. Emphasis is placed on the comprehension of mechanical behaviour and not on computational details. The methods presented should be of help to young dam engineers who lack practical experience.

All 26 fully worked numerical examples concern dams whose actual behaviour is known from *in situ* measurements. Comparisons of measurements and approximate analyses show how accurately the future behaviour of a dam can be predicted. The comparisons also show how small is the clearance which can be realized with a so-called 'exact' analysis.

To help the student master the subject, three fully worked preliminary design examples are presented in chapter 6 of this book.

Author's biography

Max A. M. Herzog, born in Czechoslovakia in 1926, graduated from the Graz Institute of Technology in Austria in 1951 with a Diploma in Civil Engineering. After working for a construction company in Austria, he gained his doctor's degree with a thesis on the analysis of arch dams by shell theory in 1956.

Later, Dr Herzog established an extensive practice as a structural and geotechnical consultant for bridges, buildings, dams, industrial and hydraulic structures, landslides and tunnels in Austria, France, Germany, Italy, Switzerland, the United Kingdom and the USA.

Dr Herzog has published more than 330 papers and 12 books. In 1985, he received the Gold Medal of the Institution of Engineers (India).

Nomenclature

A	sectional area
a	distance between buttresses or developed length of dam base or dimension of lever arm
a_E	response acceleration of dam
B	width of dam section or base thickness of gravity dam
B_0	crest length of arch dam
B_m	arch length at midheight of dam
b	head width of buttress or dimension
b_0	thickness of buttress
b_c	width of core
b_E	ground acceleration due to earthquake
b_f	width of filter
b_s	width of shell
C	adhesion
C_s	seismic coefficient
c	cohesion or coefficient of adhesion or eccentricity
D	ratio of interior damping
d	head thickness of buttress
E	Young's modulus of elasticity or deformation (E_c of concrete, E_M of model material, E_r of bedrock) or earth pressure or equivalent static earthquake load (E_h horizontal and E_v vertical component)
E_{dyn}	dynamic modulus of elasticity
e	eccentricity
f	ductility or freeboard or frequency of vibration (f_1 natural frequency, f_E earthquake frequency)
G	shear modulus or deadload
G_B	deadload of buttress
G_D	deadload of a 1 m thick slice of slip body
G_{dyn}	dynamic shear modulus
G_S	deadload of shell or buttress socle
G_V	deadload of vault

g	Earth's gravitation (9·8 m/s^2) or deadload (g_n normal and g_t tangential component)
H	height
h	actual height of dam above soil surface
I	moment of inertia
K	coefficient ($\bar{K}$ of rigid fixity, K^0 of hinged support)
k	coefficient of permeability or auxiliary value in numerical integration by Simpson's rule
L	length of dam crest
L_s	length of reservoir surface
M	bending moment (M_x in meridional and M_y in ring direction, M_{yA} at abutment and M_{yC} at crown) or Richter's magnitude of an earthquake
M_{1-4}	multipliers covering different influences on the seismic coefficient
m	ratio of uplift to hydrostatic pressure
N	normal or direct force (N_x in meridional and N_y in ring direction)
n	porosity or inclination of the dam's slope or scale of model
p	hydrostatic pressure (p_y portion transmitted in ring direction to the abutments)
p_g	overburden pressure
p_v	vertical pressure
Q	shear force (Q_x in meridional direction)
q_z	fictional deadload of dam acting in horizontal (transverse) direction
R	radius (R_e exterior radius) or resultant force
R_c	compressive strength of concrete
R_{ct}	tensile strength of concrete
S	section modulus (S_d of downstream and S_u of upstream face of dam)
s	safety or coefficient of safety
T	variation of temperature or depth of water or tangential force
T_1	natural period of vibration
T_b	basic value of response spectrum
ΔT	difference of temperature
t	thickness of dam (t_0 at crown, t_m at midheight and t_u at base)
U	uplift
u	settlement (u_1 during construction, u_2 due to consolidation) or vertical displacement or pore pressure
V	volume of concrete or dam body
v	horizontal displacement in longitudinal direction

W	hydrostatic pressure (W_h horizontal and W_v vertical component)
w	radial or horizontal (w_z) displacement in transverse direction
w_1	crest displacement due to fictional deadload acting in horizontal (transverse) direction
x	meridional coordinate or depth of water
y	ring coordinate
z	vertical coordinate
z_0	critical or limit depth
α	average inclination of slip circle or inclination of resultant force
β	inclination of valley flank
γ	density (of dam body)
γ_c	density of concrete
γ_w	density of water
γ_M	ratio of densities of model load and water (e.g. for mercury $\gamma_M = 13{\cdot}6$)
δ	angle of wall friction
ε	strain or deviation from vertical or angle between resultant force and inclined valley flank
ε_c	strain of concrete
ε_s	specific shrinkage
η	efficiency of arch dam
λ	characteristic length of circular cylindrical shell
λ_0	coefficient of earth pressure at rest
μ	Poisson's ratio (for concrete $\mu = \frac{1}{6}$)
σ	normal or direct stress
$\sigma_{1,2}$	principal stresses
σ_c	compressive concrete stress
σ_h	horizontal tensile stress in buttress
σ_z	tensile surface stress of valley floor
τ	shear stress
ϕ	angle of friction
ω	coefficient of thermal expansion or angle in plan

Contents

1. Historical development of dam analysis

Even though the history of dam construction has been covered recently in the literature,[1] such a history must also be included in a book on dam analysis because historical developments[2] help better to understand present-day methods.

1.1. Earth and rockfill dams

Constructed six thousand years ago (Waja Dam in Jordania), the oldest dams were made of earth and rockfill because of their simple technology and favourable costs. Different dam failures — slope slips (1842 Cercey Dam in Burgundy, France), underground leakage (1802 Estrecho de Rientes, Spain) and overtopping (1889 South Fork of the Conemaugh River near Johnstown, Pennsylvania, USA) — have demonstrated the lack of adequate design methods for embankment dams and distinguished the different types of construction already in use in the 19th Century such as hydraulic or compacted fill of a uniform or zoned material with a central core or impervious surface of clay, asphalt or concrete. It is therefore rather surprising to learn that the US Bureau of Reclamation did not begin research into the problems of earth and rockfill dams with greater intensity before 1936.

The major problem is the possibility of sliding of dam slopes, especially the upstream slope as a result of rapid lowering of the reservoir level. Simple sliding planes of straight lines and circles (Fig. 1), for which the sliding body is considered as undeformable (after Fellenius[3]), must be taken into account, as should more complicated geometries with deformable sliding bodies (Fig. 2). *Coulomb*'s law of friction (of 1773) is still the valid criterion of sliding as long as the influence of pore pressure in the dam body with the help of *Terzaghi*'s effective stress is considered. In earlier times only the two-dimensional or plane case was analysed. But since the catastrophic Monte Toc Slip into the full Vaiont reservoir in 1963 the three-dimensional or spatial case[6,7] has been considered. *Herzog*[8] extended the plane strip method of *Fellenius*[3] by introducing differential columns of the slip body and integrating over two orthogonal directions (Fig. 3). This calculation method yields recog-

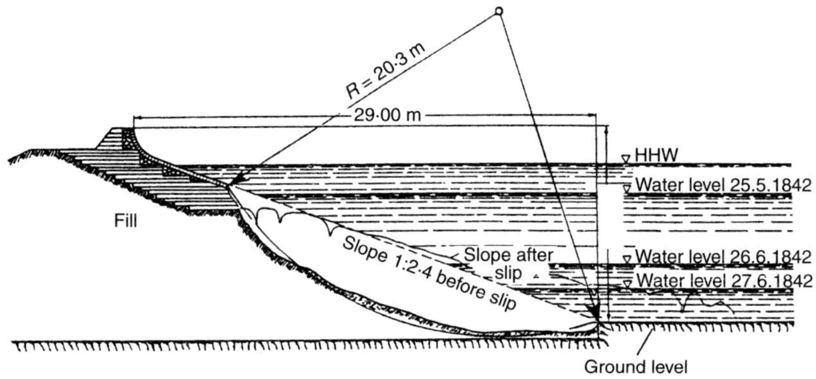

Fig. 1. Slip of the upstream slope of Cercey earth dam in 1842[4]

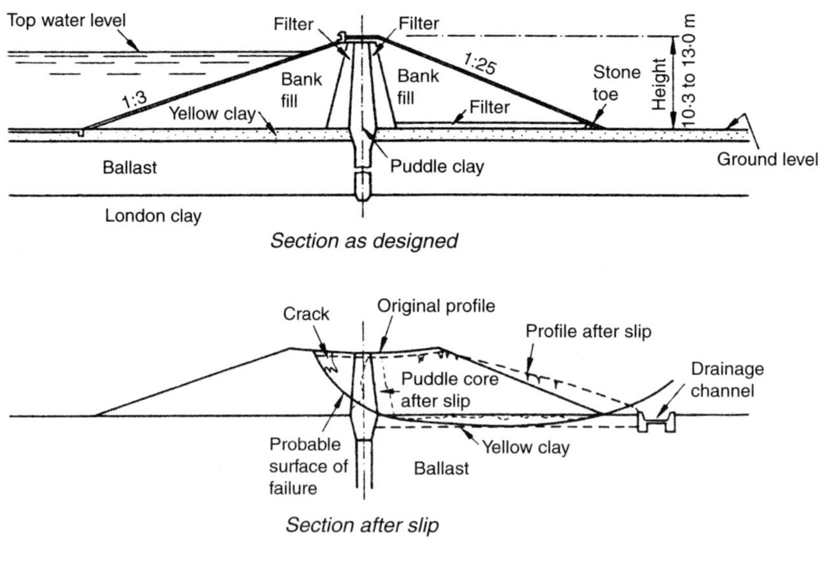

Fig. 2. Non-circular slip of the downstream slope of Lea River Dam near Chingford, Essex, England, in July 1937[5]

nition of where the slip starts and when the point of no return is reached.

The second problem is vertical settlements and horizontal displacement of the dam. Construction settlements that occur during the fill phase must be distinguished from consolidation settlements that start after dam construction (Fig. 4),[9] both for the dam body and the com-

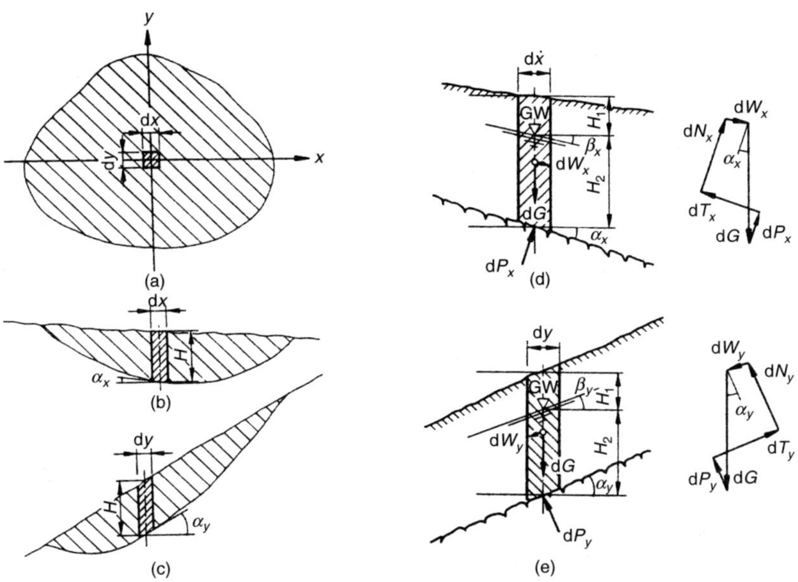

Fig. 3. Differential column element of a spatial slip body with acting forces[8]

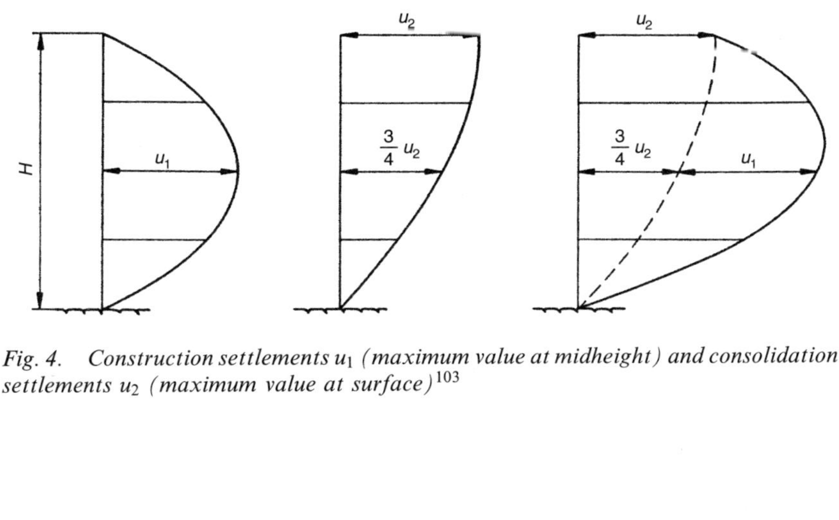

Fig. 4. Construction settlements u_1 (maximum value at midheight) and consolidation settlements u_2 (maximum value at surface)[103]

pressible underground above the bottom rock. The prediction of dam deformations depends on the values of deformation moduli for both the dam body and the underground. They are difficult to estimate in advance, as exemplified by the case of the 153 m high Gepatsch Dam in the Tyrol, Austria.[10,11] The surface displacements of embankment dams, especially crest displacements (Fig. 5), are a function of the reservoir level, dam deformation and the underground situation.[12] Above rock only the deformation of the dam body is important. In general, the water pressure on the dam's central core surpasses the difference of earth pressures of the downstream and upstream shells due to their different effective densities (downstream shell not submerged and upstream shell submerged). Above a compressible underground the influence of reservoir load can preponderate. Then, the crest of the dam can move upstream for a full reservoir. Above a sloped foundation rock the different thicknesses of the compressible valley fill can cause additional crest displacements.

Pore pressures in the dam body, which have to be compared with interior stresses due to deadweight and external water pressure,[13] form the third problem. But here common sense must prevail. There is no sense in considering pore pressure in a rockfill dam of high porosity, whereas in homogeneous earth dams and in the clay core of rockfill dams it is vital to keep the pore pressure below the overburden pressure. In the central core of a rockfill dam vertical deadload stresses can be lowered (by the so-called 'silo' action[9]) to such a value that the then preponderant pore pressure causes the feared hydraulic fracture. The arising cracks in the dam core then not only allow increased seepage (an alarm signal of prime importance) but they also render possible internal erosion with all its consequences and which cannot be stopped once started.

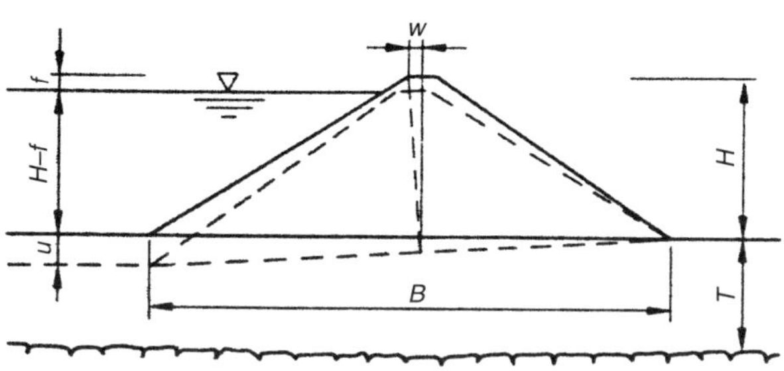

Fig. 5. Crest displacement of a well compacted dam upon a compressible underground due to reservoir load[12]

The fourth problem is the possibility of sliding of the dam as a whole, as experienced in the earthquake of 1925 in Santa Barbara, California, which led to the liquefaction of the underground consisting of silty sand.[14]

Using the method of finite elements,[15–17] stresses in and displacements of earth and rockfill dams (Fig. 6) can now be predicted, even with consideration of seepage both in the dam body and in the underground at any point of the dam.[18] As the cost of such three-dimensional analyses is still large, the use of simple approximate methods that can cope with slope stability, settlements and displacements, silo action of the core and interior stresses is recommended for design studies and preliminary designs. In the case of narrow valleys[19,20] the portion of water load carried to the valley flanks (Fig. 7) can be quantified sufficiently accurately by the so-called 'arch' actions in both longitudinal and transverse directions of the valley.[9]

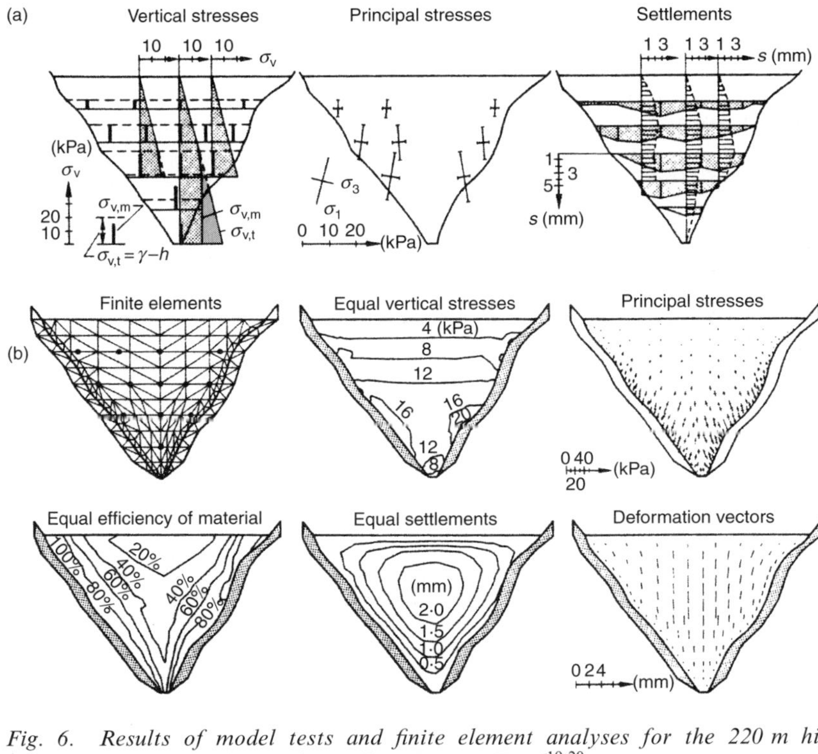

Fig. 6. Results of model tests and finite element analyses for the 220 m high Dabaklamm Dam in the Dorfer Valley, Eastern Tyrol[19,20]

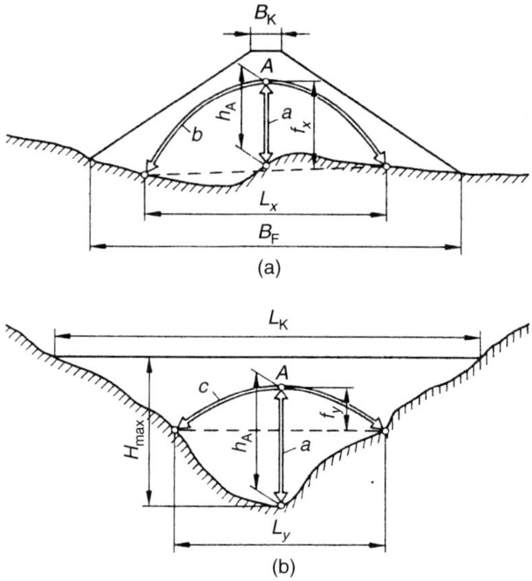

Fig. 7. Load-carrying action of an embarkment dam in a narrow valley subjected to deadload (a) *in longitudinal and* (b) *in transverse direction of the valley*[9]

1.2. Gravity dams

Until the rupture of the Bouzey Dam near Epinal in France (Fig. 8) in 1895,[21] it was customary to check the highest cross-section of the dam (assumed to be rigid) only for tilting about the toe, for the safety against sliding of the dam base, and for the vertical interior stresses. When *Rankine*[22] was asked for advice on the design of Periyar Dam in India in 1870, he mentioned two additional design criteria for gravity dams. First, the normal stress of the inclined downstream face is bigger than its vertical component. Second, the resultant force must stay within the middle third of the cross-section if tension on the dam surface is to be prevented. It was not until 1898 that *Lévy*[23] reported equations of interior stresses in a gravity dam considered as an elastic wedge. In 1905 and again in 1907 – 1908 *Wilson* and *Gore*[24,25] reported tests with rubber models to clarify the question of interior stresses in gravity dams (Fig. 9). In 1909 *Richardson*[26] showed how the plane stress equation of the theory of elasticity for a gravity dam can be solved with the method of finite differences. It is a pity that his solution was not noticed by dam designers at that time.

In 1889 *Kiel*[27] mentioned that reservoir water can enter a dam body in cracks and open construction joints only when the water pressure is

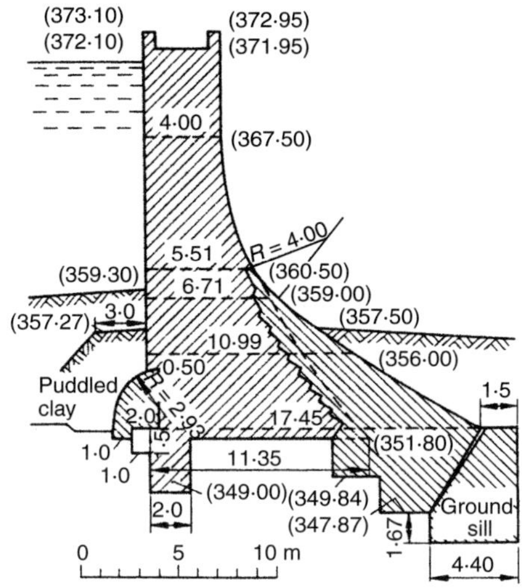

Fig. 8. Cross-section of Bouzey gravity dam near Epinal, France, built from 1877 to 1882 and collapsed in 1895[21]

larger than the vertical compressive stress. In 1898 *Lieckfeldt*[28] formulated a proof of stresses considering joint water pressure. The problem of percolation — considering dams as porous media — was taken up by *Fillunger*[29] in 1913. His theoretical views were vigorously contested by *Terzaghi*[30] and in 1934 he committed suicide. In 1938 *Heinrich*[31] showed that the correct solution of the percolation problem must start from a two-phase system. In 1936 *Brahtz*[32] reported the application of both *Airy*'s stress function and photoelasticity to the stress analysis of gravity dams, considering different elastic properties for both dam and underground. Different *Young*'s moduli of elasticity for dam and underground were taken into account in the mathematical theory of functions by *Tölke*[33] in 1938 (Fig. 10). The enormous amount of computation involved was avoided in 1947 by *Zienkiewicz*[34] who solved the system of difference equations with the help of the method of relaxation, devised by *Southwell*[35] in 1940. When designing the 285 m high gravity dam Grande Dixence in Switzerland, *Stucky*[36] studied the influence of rock elasticity with the help of model tests as had been done for the 221 m high Hoover Dam in the USA about 20 years previously. Of particular importance to the proof of stability of gravity dams is the assumed distribution of water pressure at the dam base (Fig. 11[37]) which depends both on the position of the grout curtain and on the efficiency of the dam base drainage. In 1965 *Rescher*[38] reported a detailed analysis of the main

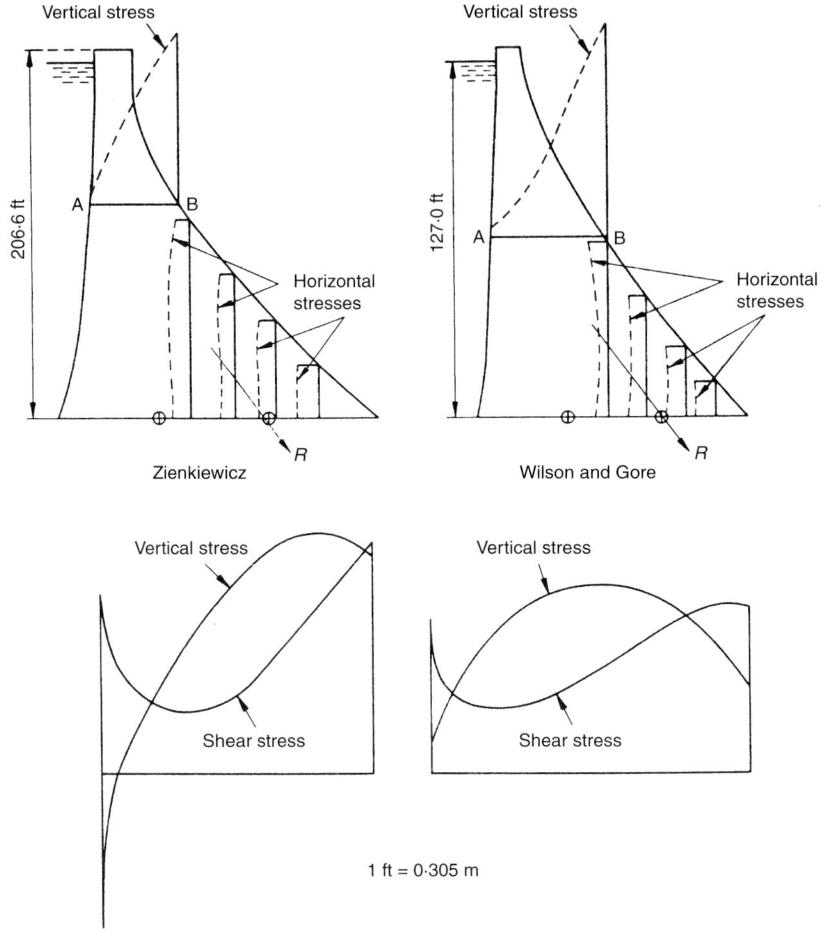

Fig. 9. Stresses in a gravity dam according to tests with rubber models[24,25] *and to finite difference analysis*[34]

cross-section of the 285 m high Grande Dixence Dam as a two-dimensional problem, even considering its construction in steps.

Since the first publication of the finite element method by *Turner et al.*[39] in 1956 (originally developed for the analysis of aeroplane structures) it was only four years until the cross-section of a gravity dam was analysed by this new method in 1960.[40] In 1975 *Campbell* and *Zienkiewicz*[41] analysed the three-dimensional load-carrying action of a gravity dam (Fig. 12) with the finite element method. In 1989 *Herzog*,[42] on the basis of long-term measurements in

Switzerland (74 years for the 112 m high Schraeh Dam in the Waegi Valley, Canton Schwyz and 40 years for the 285 m high Grande Dixence Dam in the Hérémence Valley, Canton Valais), succeeded in proving that the spatial action of straight gravity dams in narrow

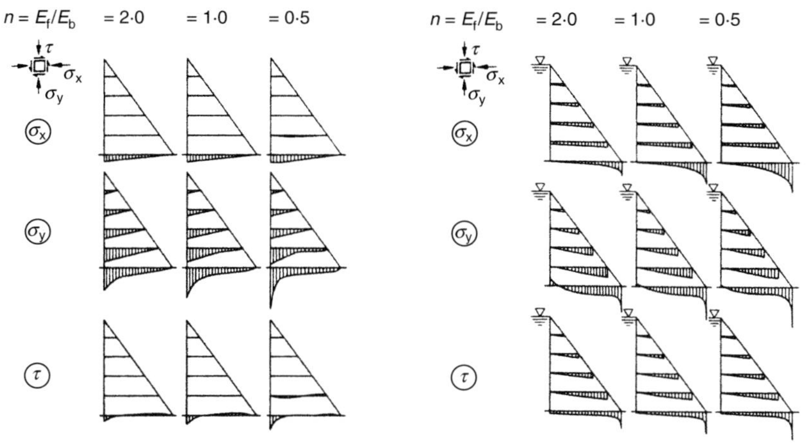

Fig. 10. Influence of the ratio of Young's moduli of elasticity of concrete and bedrock on stresses in gravity dams[33]

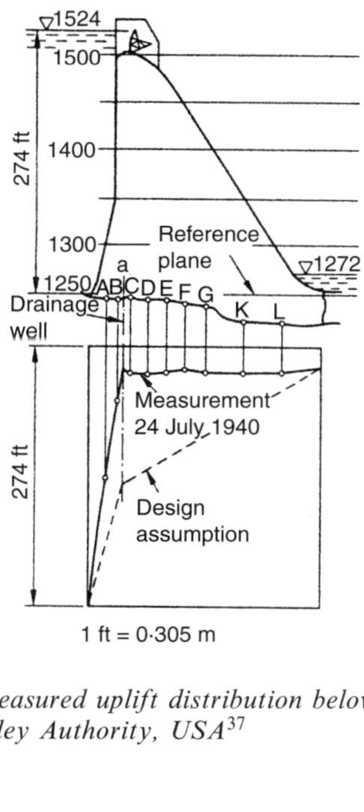

Fig. 11. Predicted and measured uplift distribution below the 89 m high Hiwassee Dam of the Tennessee Valley Authority, USA[37]

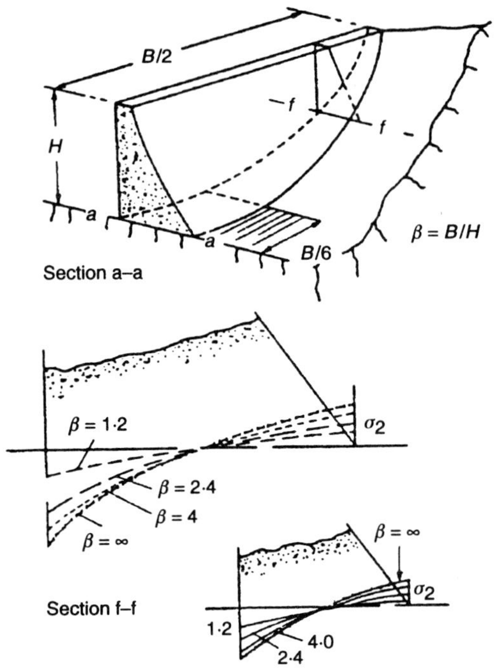

Fig. 12. Influence of the ratio of crest length to dam height on the base stresses of gravity dams subjected to deadload and hydrostatic pressure[41]

valleys exists beyond doubt. It can be accounted for with a simple grid analysis considering only one cantilever and the inability of vertical construction joints to carry tension (Fig. 13). Neither the dissipation of the cement's heat of hydration nor widening of the valley due to the reservoir load put the three-dimensional load-carrying action in question.

1.3. Arch dams

Analyses of arch dams were performed until the 1920s (e.g. in 1925 for the 113 m high Pacoima Dam in California) with help of *Mariotte*'s (1620 – 1686) well known ring or tube formula. In 1889, 24 years before the work of *Ritter*[43] who is considered as the originator of the method in Europe, *Vischer* and *Wagoner*[44] described the adjustment of radial displacements of an arch dam by assuming several horizontal arches and only one (crown) cantilever. This method was first applied in 1904 to the analysis of the 67 m high Pathfinder Dam and the 99 m high Buffalo Bill Dam (Fig. 14) in

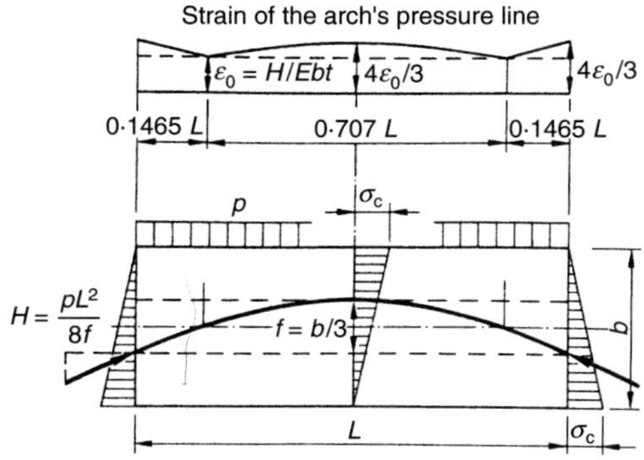

Fig. 13. Horizontal section of a gravity dam in a narrow valley acting as a tension-free beam with the strain distribution in the pressure line of an equivalent arch[42]

the USA.[45] In order to utilize better the employed concrete in his design of the 52 m high Salmon Creek Dam in Alaska (Fig. 15) in 1913 – 1914, *Jorgensen*,[46] following an earlier proposal by *Eastwood* [47] in 1910, departed from the constant radius type of arch dam and realized the first constant angle dam. Since then constant angle dams have been the dominant dam shape in not too wide valleys. Nevertheless, there have been a few retrogressions: the 114 m high Spitallamm Dam in Switzerland built in 1926 – 1932, the 180 m high Tignes Dam in France in 1949 – 1953 and the 245 m high Sayanoshushensk Dam in Siberia in 1970 – 1988. The 55 m high Montsalvens Dam built in Switzerland (Canton Fribourg) in 1919 – 1920 was analysed by *Stucky*[48] using a radial adjustment of four arches and nine cantilevers (Fig. 16). This grid method was extended in the USA from 1923 with tangential and twist adjustments, and presented in 1929 by *Howell* and *Jaquith*[49] as the 'trial load method', because the division of loads had to be found by successive trials. By 1925 *Vogt*[50] had shown that the displacements of foundation rock could be predicted approximately using three formulae derived from *Boussinesq*'s theory of the elastic semi-space.[51] The trial load method had already been checked in 1926 with the help of the Stevenson Creek Test Dam (Fig. 17), an 18 m high and only 61 cm thick experimental arch dam[52] near Fresno in California. The trial load method underwent its first application on a truly grand scale in Houk's analysis of the 221 m high Hoover Dam (Fig. 18) in the USA.[53] The results of the extraordinarily voluminous computations were checked with a model test (scale 1 in 240) at the University of Colorado at Boulder (Fig. 19).[54]

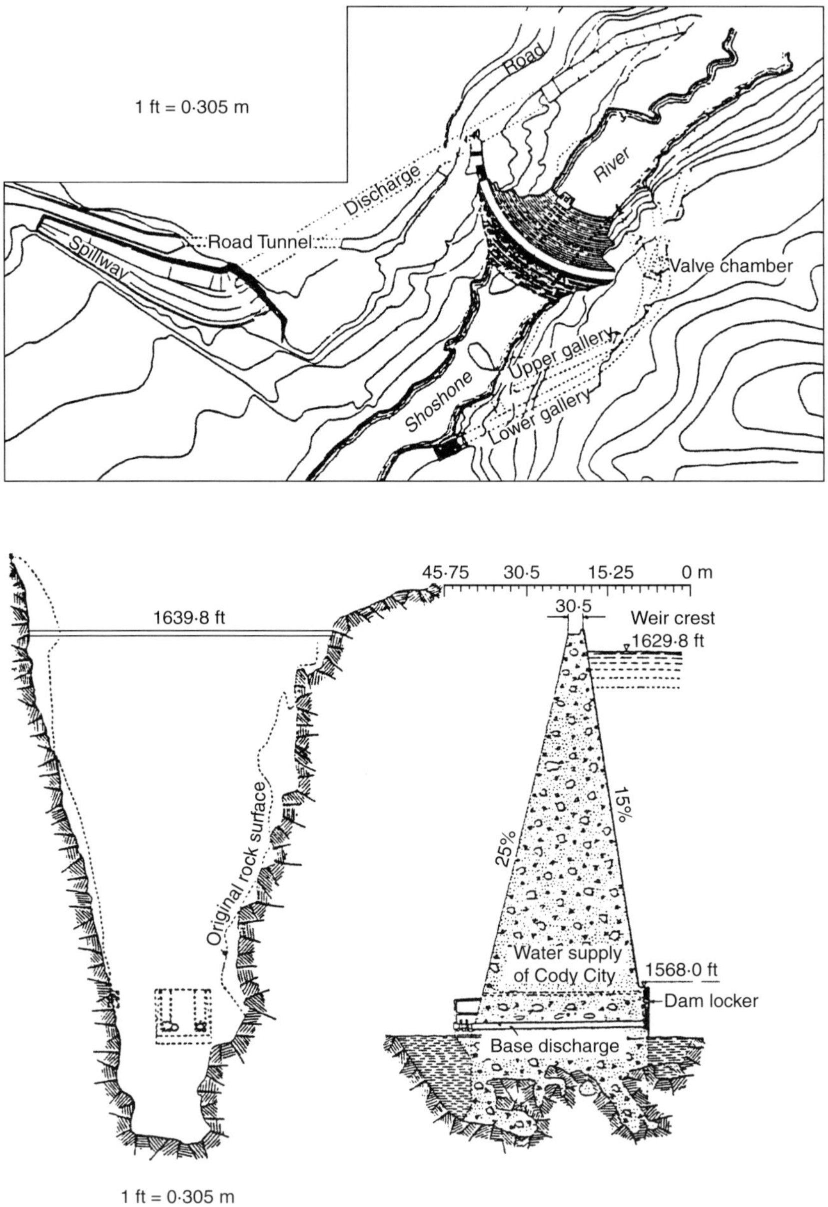

Fig. 14. The 99 m high constant radius arch dam Buffalo Bill near Cody, Wyoming USA[45]

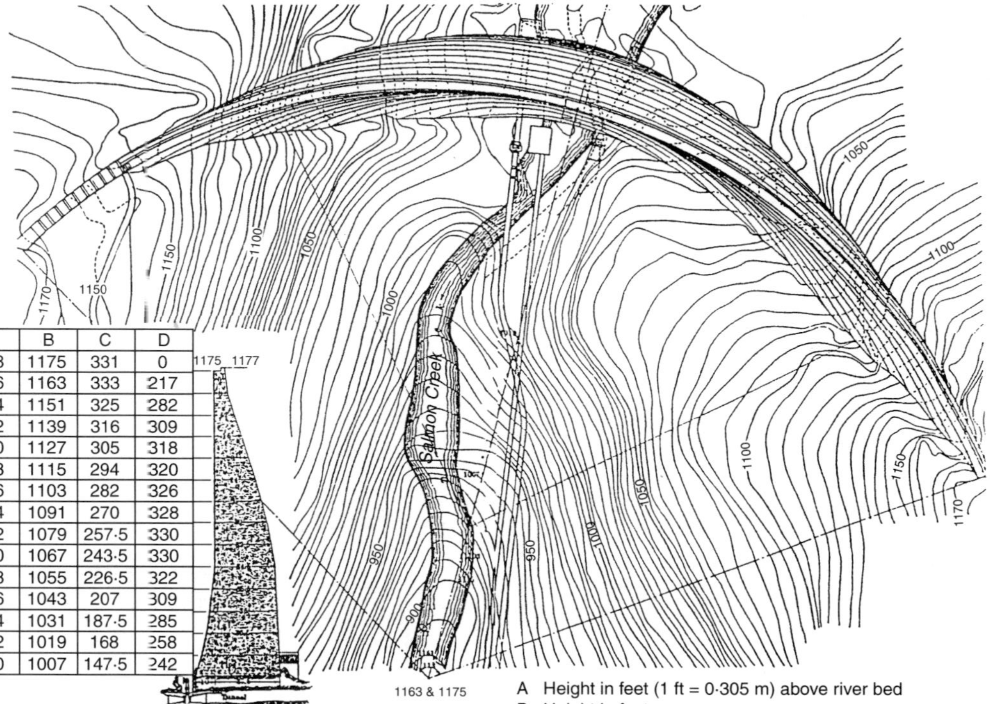

A	B	C	D
168	1175	331	0
156	1163	333	217
144	1151	325	282
132	1139	316	309
120	1127	305	318
108	1115	294	320
96	1103	282	326
84	1091	270	328
72	1079	257·5	330
60	1067	243·5	330
48	1055	226·5	322
36	1043	207	309
24	1031	187·5	285
12	1019	168	258
0	1007	147·5	242

A Height in feet (1 ft = 0·305 m) above river bed
B Height in feet
C Upstream radius in feet
D Ring stress in psi (1 psi = 0·007 MN/m^2)

Fig. 15. The 52 m high constant angle arch dam on the Salmon Creek in Alaska[46]

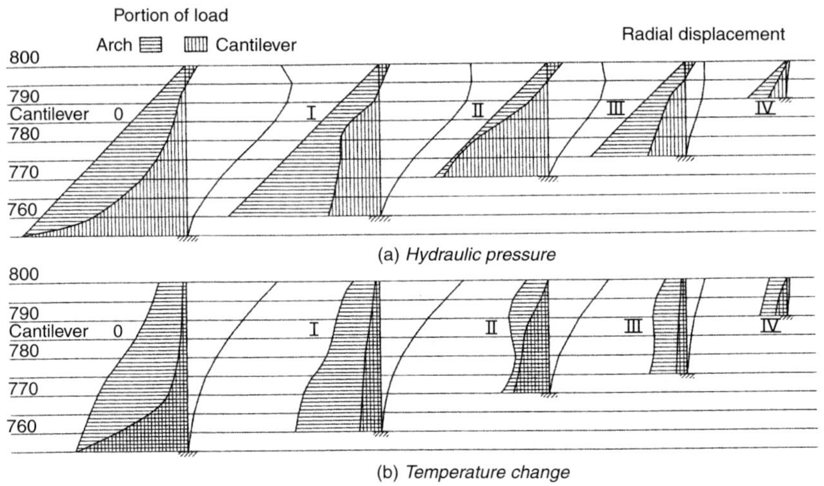

Fig. 16. Radial displacements and load distribution of the 55 m high constant angle arch dam Montsalvens on the Jogne River in Switzerland[48]

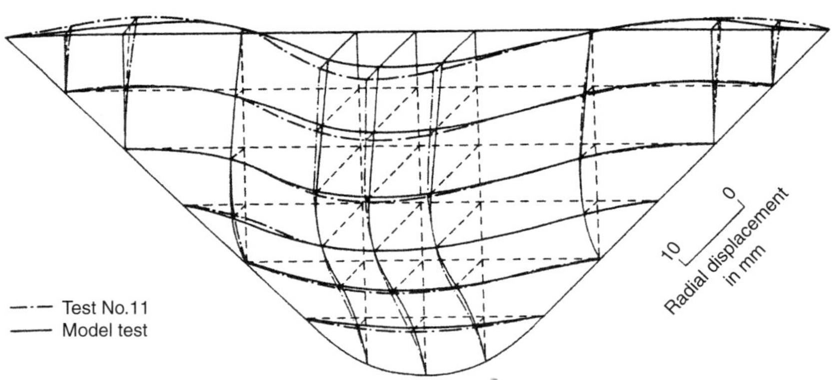

Fig. 17. Radial displacements of the 18·3 m high constant radius arch dam on the Stevenson Creek near Fresno in California[52]

Because of the enormous cost of computations, the trial load method did not find partisans in Europe. In France and Italy the analysis of arch dams with the help of horizontal elastic arches was even prescribed by the official dam regulations. It was not until 1948 that *Jurecka*[55] showed in his doctor's thesis how the grid problem could be solved without a laborious trial and error procedure by an algebraic system of elasticity equations, resulting in the division

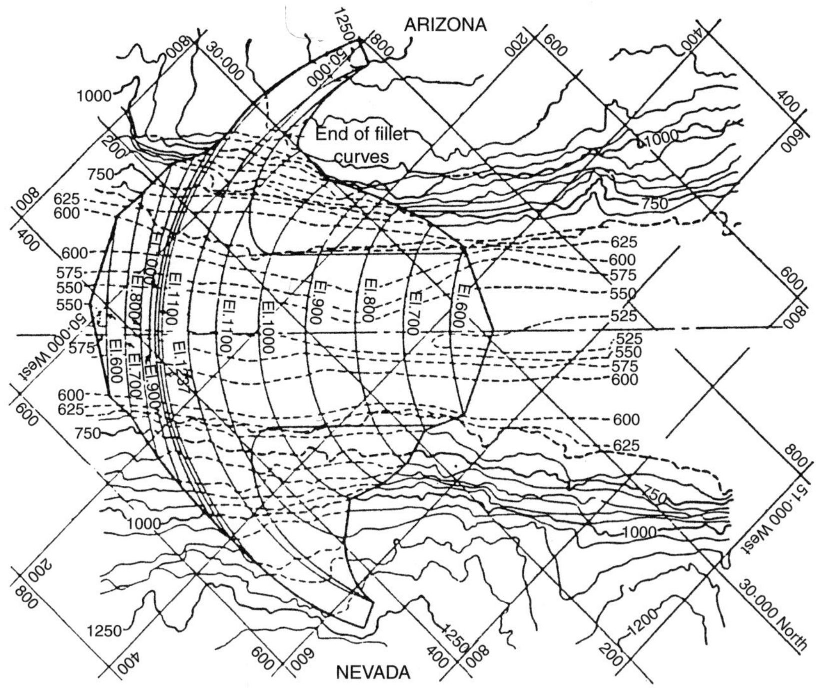

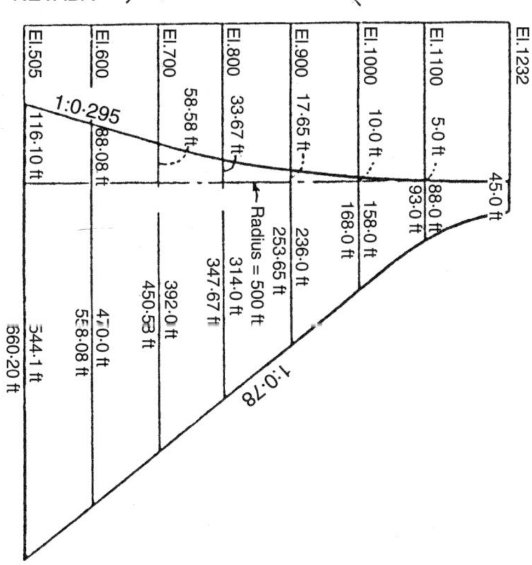

El.	*R* in feet (1 ft = 0.305 m)				Arch length in feet	Angle of aperture in degrees
	Up-stream	Down-stream	Haunch Arizona	Haunch Nevada		
1232	500	455			1167	140
1100	505	412	215	262	896	112
1000	520	352	229	249	704	92.5
900	593	339	191	217	572	70.5
800	709	361	143	166	478	52.5
700	834	383	95	110	369	35.0
600	963	405	46	53	180	30.0

Fig. 18. The 221 m high Hoover Dam on the Colorado River in the Black Canyon between Arizona and Nevada, USA[53]

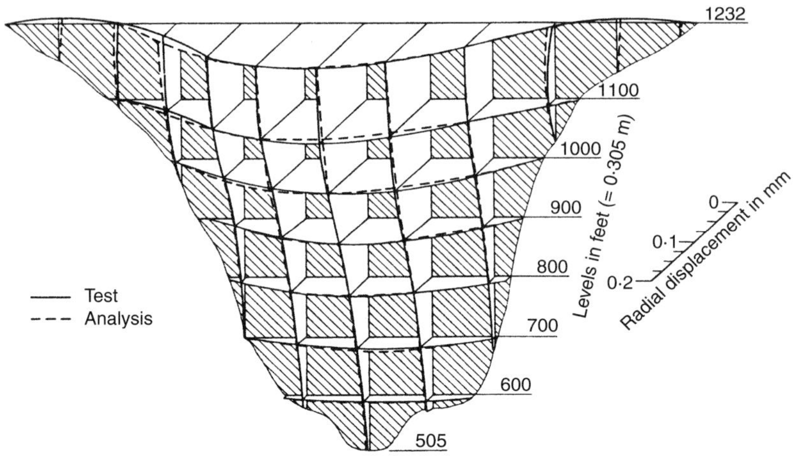

Fig. 19. Radial displacements of the plaster-kieselgur model of Hoover Dam to the scale of 1 in 240 according to measurement and prediction by the trial load method[54]

of loads between horizontal arches and vertical cantilevers. Later, the introduction of electronic data processing allowed consideration of up to six different adjustments.[56] Only this complete adjustment method was able to predict[57] the otherwise unpredictable displacements (Fig. 20) of the 90 m high and not more than 1·5 to 4·2 m thick Tolla Dam in Corsica.

Parallel to the grid methods, the development of arch dam analysis with the help of shell theory took place. In 1919 *Smith*[58] reported an application of tank theory to the problem of valley flanks in Australia. In 1938 *Toelke*[33] tried, also not very convincingly, to solve this problem with a four-step procedure. It was not until 1951 that *Tschech* and *Jaburek*[59] introduced a self-evident artifice to cope with the fixity of arch dams in their abutments. The integration of three partial differential equations as required by the shell theory is a very considerable undertaking. This was demonstrated by *Rescher*[60] in 1951 for the simplest case of a circular cylindrical shell of constant thickness with help of the already mentioned relaxation method.[35] In 1955 *Lombardi*[61] carried out the integration of the partial differential equations of a circular cylindrical shell of variable thickness using suitable polynomials. In 1956 *Allen et al.*[62] reported the application of finite differences to a three-dimensional analysis of the 117 m high constant radius Dokan Dam in Iraq. Despite application of the relaxation method, these computations took three years (1952 – 1955). In 1970 *Gilg*[63] reported analysis of the 130 m high

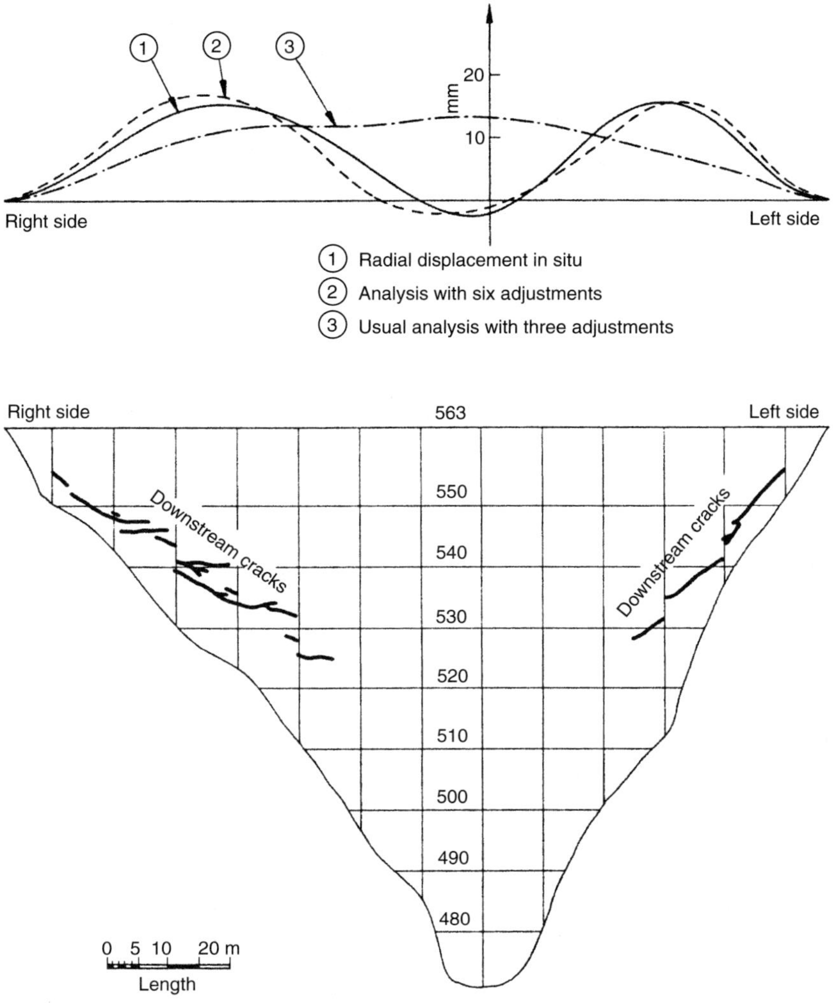

Fig. 20. Crest displacements of the 90 m high and very thin Tolla Dam in Corsica after in situ *measurement and prediction by six adjustments*[57]

double-curvature Punt dal Gall Dam (Fig. 21) in Switzerland by the theory of shells. In this work 700 equations had to be solved for as many unknown displacement components. Self-evidently, Gilg succeeded only with help of a powerful computer. A number of mathematically elegant methods to solve systems of partial differential equations do exist but none gained practical significance.

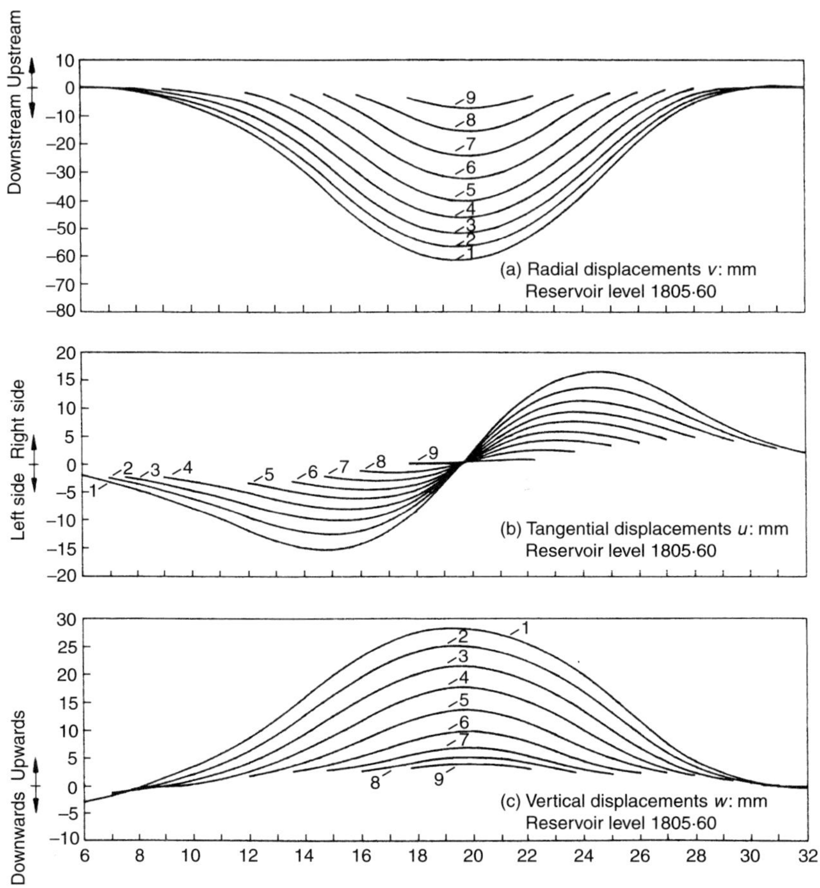

Fig. 21. Radial, tangential and vertical displacements by shell theory of 130 m high constant angle dam Punt dal Gall in the Livigno Valley, Italy[63]

In 1992 *Herzog*[64] presented an approximate method for preliminary design and checking purposes. The method was shown to be entirely adequate by the posterior analysis of some 60 arch dams. It is based on solution of the tank equation by the classic edge-perturbation method and yields, with the help of closed formulae, a conclusive picture of the stresses and displacements of an arch dam subjected to water pressure and temperature changes after only an hour of manual calculation.

The finite element method[39] had to wait for development of the matrix displacement method[65] before it became possible to analyse

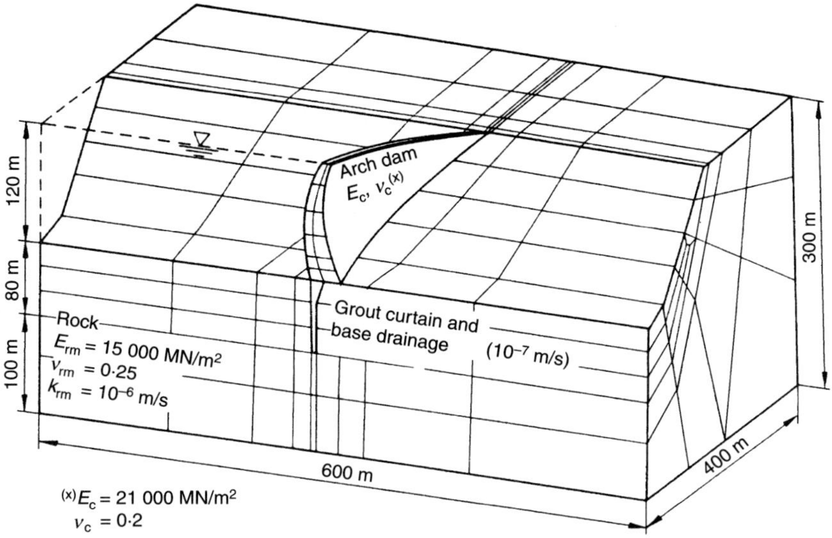

Fig. 22. Arch dam in fissured bedrock with percolation[69]

arch dams[66] with the help of isoparametric elements.[67] At present both the foundation rock[68] and percolation of the dam[69] (Fig. 22) can be accounted for. The newest developments consider the opening of construction joints and the formation of cracks.[70] In general, analysis of arch dams using the finite element method[71] requires very powerful computers. However, even as early as 1986 *Stevenson et al.*[72] reported the finite element analysis of an arch dam on a personal computer.

2. Elements of dam analysis

2.1. Loadings

Dams are subjected to the following loadings[73]

- deadweight
- water pressure
- ice pressure
- silt pressure
- temperature changes
- volume changes
- underground deformations
- earthquakes.

The action of *deadweight* can be considered as statically determinate as long as no arch dams with appreciable downstream overhangs (e.g. Couesque and Enchanet in France) are involved. In such dome-type dams the vertical construction joints must be grouted or at least closed before the individual blocks reach the crest. The density of concrete is between 24 and 26 kN/m^3. The density of earth and rockfill dams varies much more (16 to 25 kN/m^3) depending on porosity, humidity and compaction.

With a density of 10 kN/m^3 *water pressure* is the best defined loading. Ice pressure in the magnitude of 50 to 150 kN/m crest length generally plays no decisive rôle. In contrast, *silt pressure* with densities of 14 to 21 kN/m^3 should never be neglected.

Fifteen years after the first *temperature measurements* on the Salmon Creek Dam in Alaska in 1915,[74] *Houk*[75] published a diagram (Fig. 23) from which the following rule-of-thumb value

$$T = \pm 62/(t + 2{\cdot}4) \tag{1}$$

for the possible increase or decrease of concrete temperature can be obtained. If the wall thickness t is entered in metres the temperature

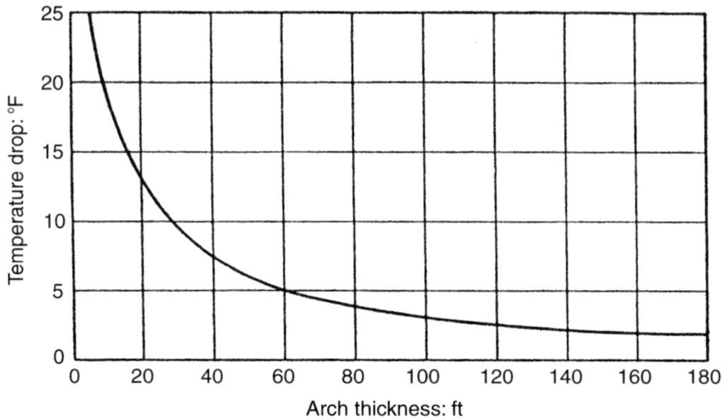

Fig. 23. Temperature variation as a function of concrete dam thickness (after Houk[75])

change T is given in degrees centigrade. In concrete dams of the usual dimensions it suffices in general to consider annual temperature changes. Daily changes need to be considered only in very thin dams such as Gage Dam in France and Tolla Dam in Corsica. In earth and rockfill dams temperature changes can be disregarded.

Although it is commonly assumed that a concrete mass does not shrink, because it practically cannot dry up, marginal *shrinkage of dam concrete* has been observed in many dams. If the few cases of autogeneous shrinkage as observed in some Portuguese dams[76] are not considered, a shrinkage of $\varepsilon_s = -3 \times 10^{-5}$ will usually suffice.[77] Nevertheless, a shrinkage value of $\varepsilon_s = -7 \times 10^{-5}$ was reported in the 237 m high Mauvoisin arch dam in Switzerland.[78]

Since the destruction of the low Sheffield Dam near Santa Barbara in California by an earthquake in 1925[14] this type of loading has been considered in dam analysis. At present the *maximum credible earthquake* with a return period of about 500 years is distinguished from the *operating basis earthquake* with a return period of about 50 years.[79] Recent failures in concrete dams (such as Koyna gravity dam in India in 1967[80] and Pacoima arch dam in California in 1971[81]) as well as in earth and rockfill dams (Upper and Lower Van Norman Dams near Los Angeles in 1971[82]) stress the importance of considering the loading case of earthquake.

Combinations of loadings are very important in the analysis of dams. In general three combinations[73] are considered. The first one comprises deadweight, changes of temperature, eventual grouting of construction joints (frequently performed in arch dams only) and, in the case of concrete dams, also the shrinkage of concrete. In

Fig. 24. Gleno multiple arch dam in Italy after collapse on 1 December 1923[93]

Fig. 25. St. Francis gravity dam in California after collapse on 12 March 1928 (7 days after first filling of reservoir)

the second combination water pressure on the dam and the effects of reservoir load on the valley floor and flanks have to be added. The third combination of loadings consists of the first or second one plus the operating basis earthquake. The occurrence of a maximum credible earthquake is considered as a catastrophe in which damage to the dam is accepted but rupture or collapse must be avoided and an alarm system to alert people living below the dam must be operational in the case of a dam break wave.[83]

2.2. Foundations

There are no other structures for which the underground is of such importance as dams. First, the underground must be of sufficient strength to resist the stresses induced by the dam. Second, the underground suffers deformations due to the very big reservoir

Fig. 26. Remnants of Malpasset arch dam in the Reyran Valley near Fréjus, France, after collapse on 2 December 1959[96]

load which can have repercussions on the dam.[84] Third, the underground is generally heterogeneous and is thus a cause of additional local stresses.

The ratio of the deformation moduli of rock and concrete ($0{\cdot}1 < E_r/E_c < 2{\cdot}0$) is of some importance in determining the magnitude of stresses. The dam analyst must make allowance for possible changes of this value along the dam's perimeter, e.g. Kölnbrein arch dam in Austria.[85] Deformations of the underground can be predicted rather easily for rock[86] but it is difficult to make realistic predictions for earth and sediments.[12] Stresses in the underground depend not only on the loads exercised by the dam[87] but also on details such as the position of the grout curtain.[88]

In complicated underground conditions, such as heterogeneous rock with cracks and faults as well as flowing or Artesian ground water, the dam analyst must resort to model tests[89] or numerical methods of analysis.[69]

Fig. 27. Baldwin Hills Reservoir in California after collapse of embankment dam

2.3. Observed behaviour

In order to determine the real load-carrying action of dams measures of their behaviour are indispensable. A large amount of observation data has been compiled in the *Proceedings of the International Commission on Large Dams* (ICOLD).[90]

In embankment and masonry dams the displacements of characteristic points are measured in dependence of the reservoir level and air temperature. Of special interest are failures which are basically *unintended full-size tests*. Their heuristic value cannot be over-estimated.

The following are some well known failures.

- 1802 Leakage under the 50 m high Puentes gravity dam near Lorca in Spain.[91]
- 1842 Upstream slope slip of the 18 m high earth dam at Cercey in Burgundy, France, during rapid lowering of the reservoir level.[4]
- 1889 Overtopping of an earth dam on the South Fork of the Conemaugh River near Johnstown in Pennsylvania, USA, due to an insufficient spillway capacity.[92]
- 1895 Rupture of the 24 m high Bouzey gravity dam near Epinal, France, due to a statically insufficient cross-section of this dam (Fig. 8).[21]

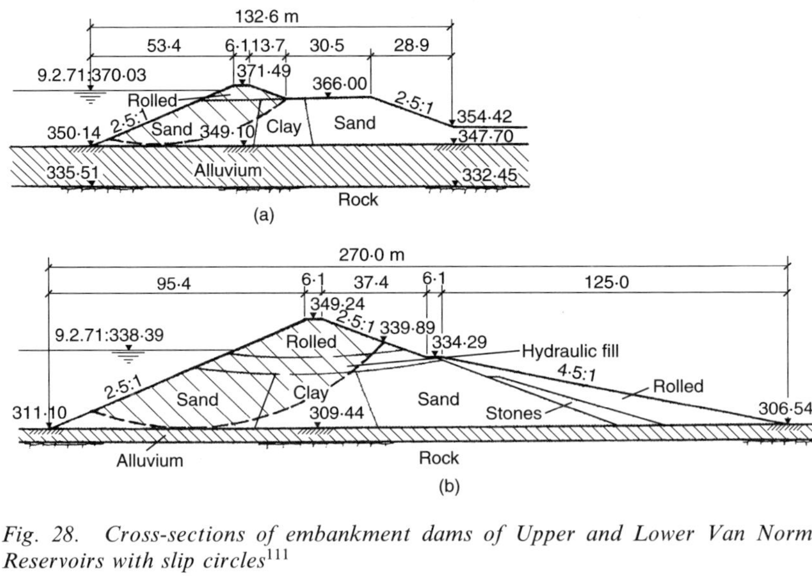

Fig. 28. Cross-sections of embankment dams of Upper and Lower Van Norman Reservoirs with slip circles[111]

- 1900 Rupture of the 21 m high gravity dam on the Colorado River near Austin, Texas, USA, by overtopping with 3·4 m depth of water.
- 1923 Rupture of the 36 m high Gleno multiple arch dam in the Bergamasc Alps, Italy, by sliding on the smooth serpentine rock which sloped downstream (Fig. 24).[93]
- 1925 Sliding of the 7·6 m high homogeneous Sheffield earth dam near Santa Barbara in California, USA, as a whole on the underground of saturated fine sand liquefacted by an earthquake.[14]
- 1928 Rupture of the 63 m high St. Francis gravity dam in California, USA, by sliding on the underground of slate and conglomerate softened by seepage (Fig. 25).[94]

Fig. 29. Weak spot in the underground of destroyed Teton Dam in Idaho, USA[100]

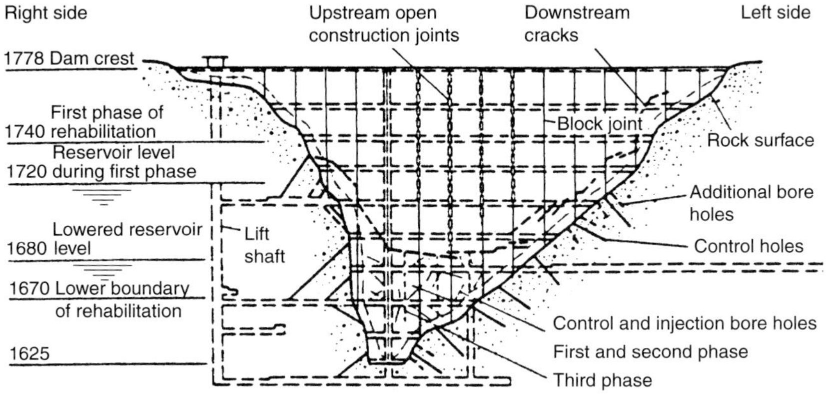

Fig. 30. Downstream view with cracks of the damaged Zeuzier constant angle arch dam in the Lienne Valley, Switzerland[101]

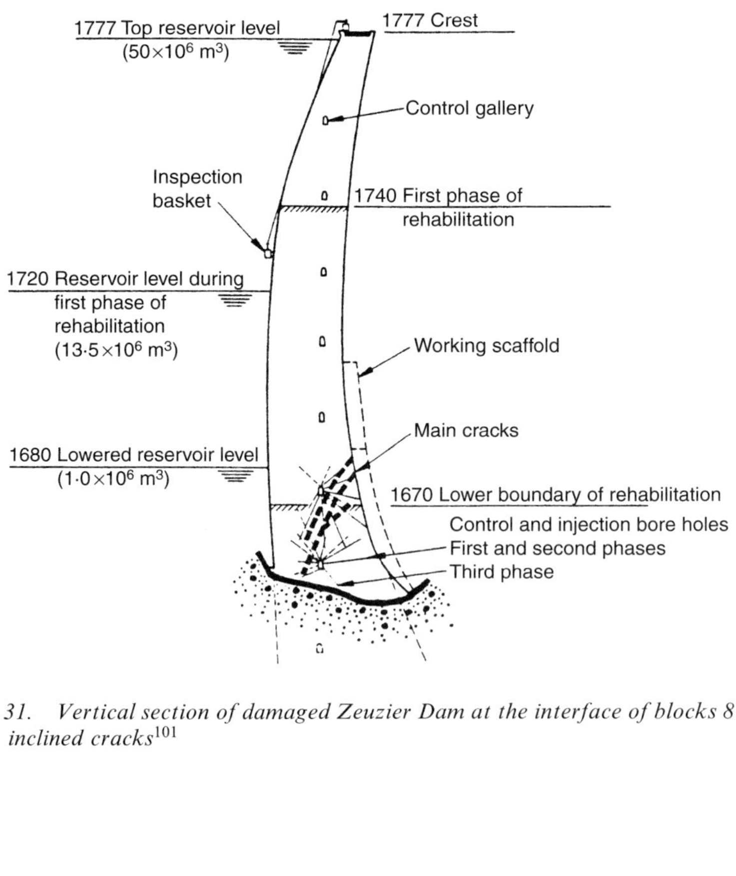

Fig. 31. Vertical section of damaged Zeuzier Dam at the interface of blocks 8 and 9 with inclined cracks[101]

- 1938 Slip of up to 300 m of the 76 m high Fort Peck hydraulic fill dam (volume 96×10^6 m^3) on the upper Missouri River in Montana, USA, due to insufficient subsoil exploration.[95]
- 1959 Rupture of the 66 m high Malpasset arch dam in the Reyran Valley near Fréjus in southern France by sliding of the dam and part of its foundation rock (gneiss) on an undetected fault plane deep below the surface (Fig. 26).[96] The possibility of upward sliding of this thin and light arch dam on the only slightly inclined valley flanks was also recently discussed.[97]

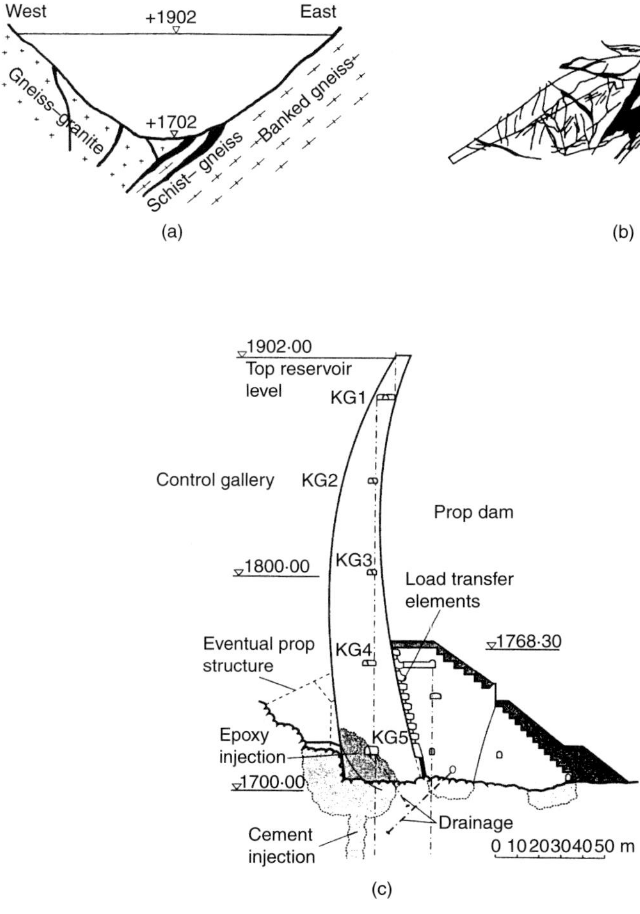

Fig. 32. Geologic situation of Kölnbrein arch dam: (a) *cross-section of the valley and* (b) *plan. (c) Crown cantilever of Kölnbrein Dam with prop dam constructed later.*[85]

- 1964 Rupture of the 47 m high Baldwin Hill Reservoir of the Los Angeles water supply in California, USA. This roller-compacted earth dam is founded on loose to moderately consolidated, young marine sediments. Subsidence of the reservoir and its surroundings were attributed to the withdrawal of oil and gas in the Inglewood Field which adjoins the reservoir on the west and south. A crack opened along an existing fault detected in 1948 during construction which created a passageway for the escaping water through the erodible foundation sediments (Fig. 27).[98]
- 1971 Slope slips on the Upper and Lower Van Norman Reservoirs of the Los Angeles water supply in California due to the San Fernando Earthquake (Fig. 28)[99] with a crest displacement of 1·5 m and a crest settlement of 0·9 m for the 38 m high dam.
- 1976 Rupture of the 120 m high Teton Dam in Idaho, USA, due to insufficient remedial treatment of a defective foundation in the rhyolite tuff (Fig. 29). The dam construction was characterized by inadequate communication between the resident engineer and the US Bureau of Reclamation in Denver, Colorado, which had to authorize any decision on the site.[100]
- 1979 Rupture by bending of the 156 m high unsymmetrical Zeuzier arch dam in Switzerland (Figs 30 and 31). Due to lowering of the ground water induced by the pilot tunnel for the planned Rawil road tunnel, a narrowing of the valley occurred with costly consequences.[101]
- 1979 Rupture by shearing of the 200 m high Kölnbrein arch dam in Carinthia, Austria due to the varying elasticity of the bedrock (Fig. 32) and the small wall thickness of the dam.[102]

The major importance of an analysis of the underground for satisfactory dam performance is clearly evident from the above mentioned failures.

3. Earth and rockfill dams

3.1. Dam types

In the early days of dam construction low dams were frequently *homogeneous* dams of moderately compacted or hydraulic fill type. However, sometimes the achieved compaction was insufficient. The simplest waterproofing of a dam can be obtained by using clay. Especially in England, the *puddled clay core* was favoured because of its simple construction. Such *heterogeneous* dams with central cores are even nowadays preferred for the highest dams (e.g. the 325 m high Rogun Dam in Tadjikistan) although the settlements of core and shells must be tuned in order to prevent the dreaded *hydraulic fracture* of the core due to the silo action of the shells. In the case of a lack of adequate particle sizes, dam cores are sometimes constructed of asphaltic concrete instead of the usual silty clay. Occasionally, dam waterproofing systems consist of thin reinforced concrete walls which reach the bedrock. In dams on alluvial subsoil the required grout curtains are produced by injecting appropriate cement or clay suspensions into the subsoil. Sometimes, a waterproof membrane is placed on or just below the surface of the upstream shell. The biggest fill volume is the 87 m high and 18 km long Syncrcede tailing dam near Ft. McMurray, Province Alberta, Canada, at 540×10^6 m^3.

3.2. Settlements

In general, only the highest cross-section of a dam is analysed in the preliminary design stage as a two-dimensional or plane case.

3.2.1. Theoretical basis

Since the first complete settlement recordings for earth and rockfill dams — for both the complete duration of the fill and consolidation phases and for the entire height of the dam and not only for its surface — it has been known that the fill and consolidation phases must be clearly distinguished if the theoretical settlement prediction is to be realistic.[103]

For the dam cross-section in Fig. 33 the *construction settlement* (the relative displacement of any point in the dam compared with the rigid bedrock due to its overburden)

$$u_1 = \frac{\gamma}{E_1}(h + t - z)z = \frac{\gamma}{E_1}[(h + t)z - z^2] \tag{2}$$

that has already occurred during the fill phase and is distributed parabolically over the height of the considered dam column (Fig. 34) has to be distinguished from the *consolidation settlement* (the settlement of any point in the dam compared with the rigid bedrock after termination of construction)

$$u_2 = \int_0^z \frac{\gamma}{E_2}(h + t - z)\,\mathrm{d}z = \frac{\gamma}{E_2}\left((h + t)z - \frac{z^2}{2}\right) \tag{3}$$

which takes place after termination of the fill phase. Its distribution over the height of the considered dam column equals a semi-parabola. It is presumed in both cases that the deformation moduli E_1 and E_2 as well as the density γ are approximately constant over the entire height of the dam column (average values).

3.2.2. Silo action

The settlements of the near-vertical central core of a rockfill dam are in general considerably reduced by the so-called silo action[104] because a portion of the core's deadweight is transferred into the adjoining shells (or shoulders) by wall friction. With the average core width b_c, the ratio of lateral to vertical pressure λ_0, and the coefficient of wall friction $\tan\delta$, *Janssen*'s silo theory[105] gives the vertical bottom

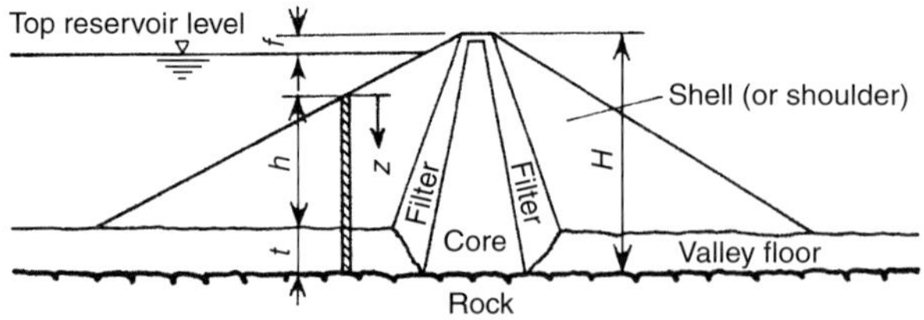

Fig. 33. Cross-section of an embankment dam with designations[103]

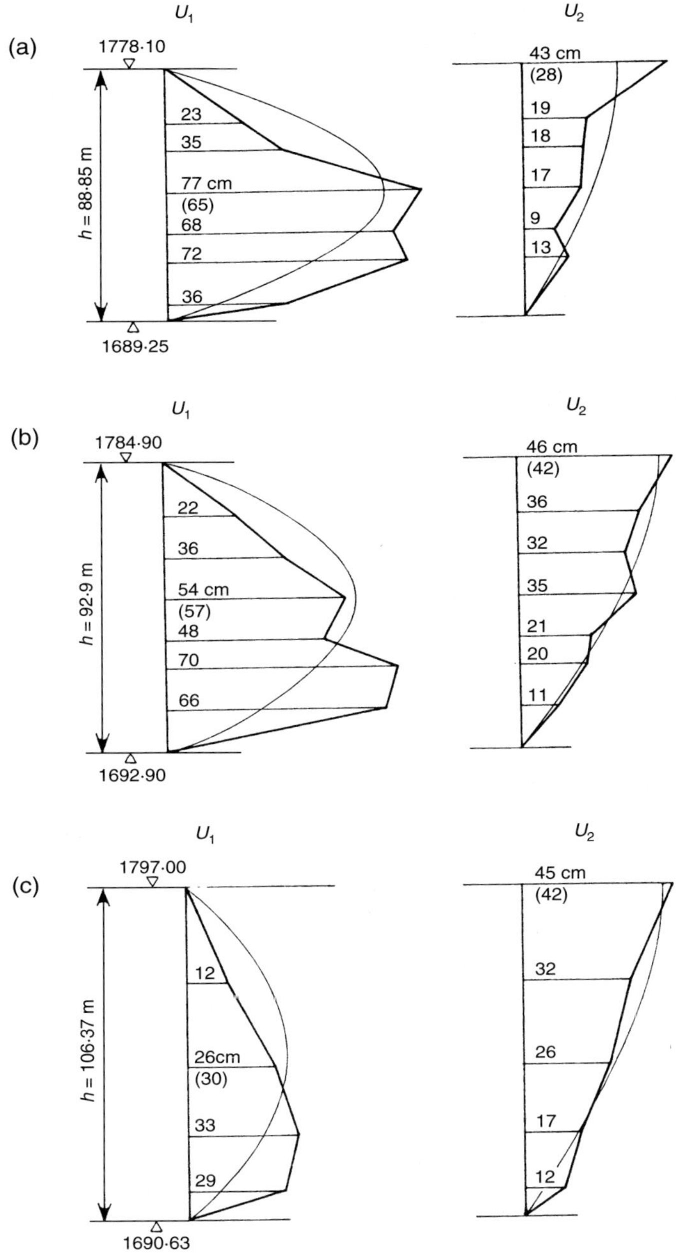

Fig. 34. Measured and theoretical distribution of settlements of Göscheneralp Dam: u_1 construction and u_2 consolidation portions of settlement respectively (a = *downstream shell SP V,* b = *upstream shell SP VI and* c = *core SP II*)[103]

pressure as

$$p_v = \frac{\gamma b_c}{2\lambda_0 \tan \delta} \tag{4}$$

Below the critical depth

$$z_0 = p_v/\gamma \tag{5}$$

the bottom pressure is smaller than the theoretical overburden pressure

$$p_g = \gamma(H - z) \tag{6}$$

The negative effect of silo action consists of hydraulic fracturing of the core if the pore pressure exceeds the vertical pressure in the dam's core.

3.2.3. Deformation moduli

An evaluation of the measured settlements of 22 big embankment dams all over the world[103] shows that there is no clear dependence of the deformation modulus on a single parameter, e.g. the density. Therefore, values of the deformation modulus required for settlement predictions can be derived only from comparisons with similar dams. The real value of the deformation modulus can only be obtained from a posterior analysis of measured settlements. As the upstream shell undergoes multiple loading and unloading sequences due to the movement of the reservoir level, its deformation modulus is always smaller than that of the only statistically stressed downstream shell.

The settlements of embankment dams are in the range of 0·2% of the dam's height (e.g. the well compacted Mica Dam in Canada[106]) to 3% of the dam's height (e.g. the poorly compacted Gepatsch Dam in the Tyrol, Austria,[107] with a porosity of 29% and additional edge breakage.)

3.3. Interior stresses and horizontal displacements

The most exact analysis of an embankment dam subjected to deadload, water pressure and earthquake can be performed with three-dimensional finite elements.[17] However, comparisons for the 244 m high Mica Dam[108] in British Columbia, Canada have shown that even an analysis of the main cross-section with two-dimensional finite elements yields results which differ only marginally. Besides these rather expensive numerical methods, elementary approximations are best suited for frequently repeated design studies. Very simple models of thought suffice for the prediction of stresses and displacements.[12] The accuracy of these approximate methods is sufficient for practical purposes.

3.3.1. Approximate analysis

The earth pressure at rest E_0 of the central core in the vertical section of the dam's longitudinal axis above the considered level (Fig. 35) is given by

$$E_0 = \frac{\gamma}{2}(H - y)^2 \lambda_0 \tag{7}$$

and the water pressure is

$$W = \frac{\gamma_w}{2}(H - y - f)^2 \tag{8}$$

Then, the maximum shear stress in the horizontal dam section is for a parabolic stress distribution

$$\tau = \frac{3(E_0 + W)}{2B_y} \tag{9}$$

Besides this shear stress the vertical overburden pressure

$$\sigma_z = \gamma(h - y) \tag{10}$$

also acts on a differential dam element. Assuming the validity of

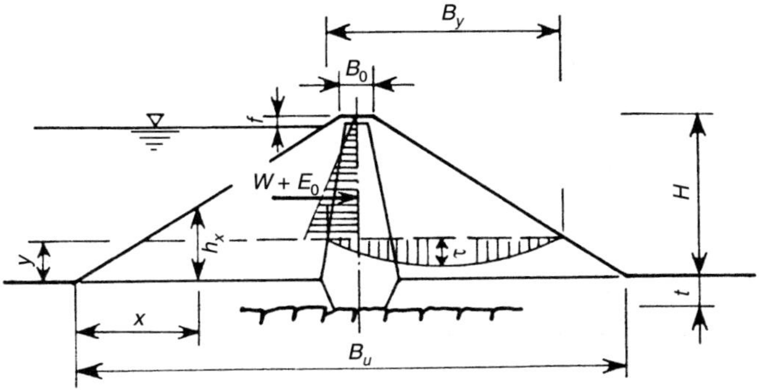

Fig. 35. Prediction of interior stresses in an embankment dam[12]

Coulomb's law of friction

$$\tau = c + \sigma \tan \phi \qquad (11)$$

the shear condition at the place of the biggest shear stress is

$$\frac{\tau}{\sigma_z} = \frac{3(E_0 + W)}{2B_y\sigma_z} \leq \tan \phi \qquad (12)$$

if the cohesion c of the dam's shell is, as usual, of no importance.

3.3.2. Shear beam method

If the real dam section is replaced by a wide wedge of identical base width and height (Fig. 36) the horizontal displacements can be easily calculated with help of the shear beam method[109] (Fig. 37). In this method only the horizontal shear deformation of the dam is accounted for[110] (note: z is counted from the crest).

(a) Deadload assumed to act horizontally

$$q_z = \gamma B_z \,\mathrm{d}z = \gamma \frac{B}{H} z \,\mathrm{d}z \qquad (13)$$

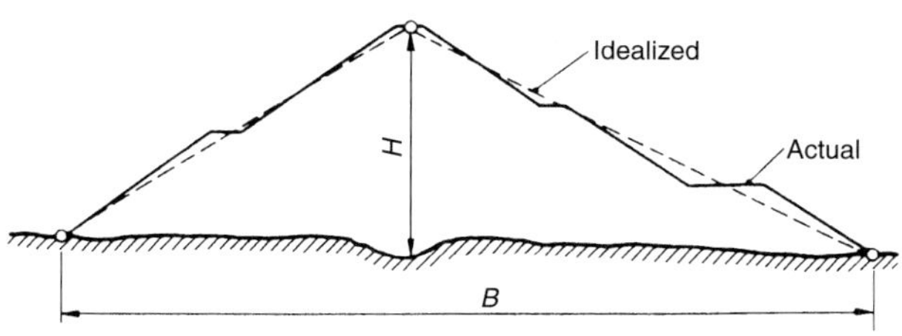

Fig. 36. Replacement of the actual embankment dam by a wide wedge[110]

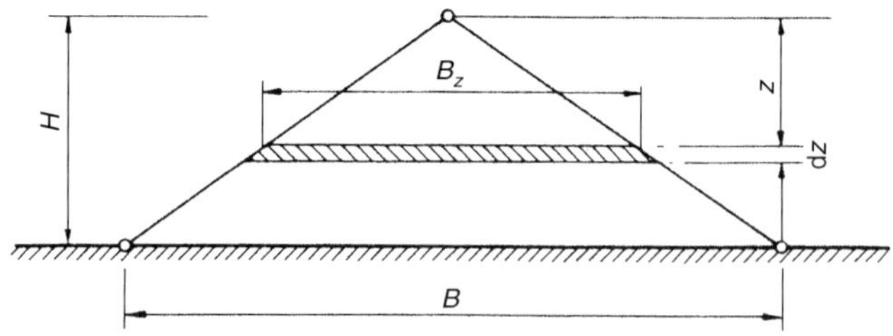

Fig. 37. Embankment dam analysed by the shear beam method[110]

(b) Horizontal shear force

$$Q_z = \int_0^z q_z \, dz = \gamma \frac{B}{H} \int_0^z z \, dz = \gamma \frac{B}{H} \frac{z^2}{2} \quad (14)$$

(c) Horizontal displacement

$$w_z = \int_0^z \frac{Q_z Q'}{GB_z} dz = \frac{\gamma}{2G} \int_0^z z \, dz = \frac{\gamma z^2}{4G} \quad (15)$$

(d) Maximum value for the dam's crest

$$\max w = \frac{\gamma H^2}{4G} \quad (16)$$

If an earthquake with horizontal response acceleration $a_{Eh} < g$ acts instead of the deadload, the horizontal crest displacement amounts to

$$\max w^E = \frac{a_{Eh}}{g} \frac{\gamma H^2}{4G} \quad (17)$$

The horizontal water pressure acting on the vertical core

$$W = \frac{\gamma_w}{2}(H - f)^2 \quad (18)$$

causes the crest displacement

$$\max w_0^W = \frac{\gamma_w (H - f)^2}{4G} \frac{H}{B} \quad (19)$$

As the submerged density of the upstream shell is diminished by the uplift

$$\gamma_a = \gamma - \gamma_w \quad (20)$$

the core's earth pressure encounters less resistance in the upstream shell than in the downstream shell. Thus, the crest displaces upstream, approximated by

$$\max w^{\Delta E} = \frac{\lambda_0 (\gamma - \gamma_w) H^3}{4GB} \quad (21)$$

The resulting crest displacement (acting downstream)

$$\max w^{W} = \max w_0^{W} - \max w^{\Delta E} \tag{22}$$

is diminished by the upstream crest displacement due to the subsoil compression by the reservoir load (Fig. 5)

$$u^{T} = \gamma_{w}(H - f)\frac{T}{E} \tag{23}$$

causing

$$w^{T} = \frac{H}{B}u^{T} \tag{24}$$

Altogether, the full reservoir can — depending on the compressibility of the subsoil — produce either a downstream or an upstream displacement of the dam's crest. Its absolute value will be small in either case.

3.3.3. Pore pressure

The influence of pore pressure can be approximated as negligible when considering the crest displacement of embankment dams with central cores.

The *effective* stress σ' in the dam body follows from the *total* stress σ by subtracting the pore pressure u as

$$\sigma' = \sigma - u \tag{25}$$

$$\tau' = c' - \sigma' \tan \phi' \tag{26}$$

and for $c' = 0$ as

$$\tau' = (\sigma - u) \tan \phi' \tag{27}$$

It is not difficult to estimate realistic approximate values of the pore pressure.

3.3.4. Direction of the displacement vector

The direction of the shell displacement approximately equals the direction of the resulting force (Fig. 38). The deviation from the vertical amounts to

$$\tan \varepsilon = \frac{E_h}{E_v + G_S} \tag{28}$$

The direction of the dam's displacement vector is frequently camouflaged by the dam's plastic deformation tendency (Fig. 39) which cannot be predicted by computation.

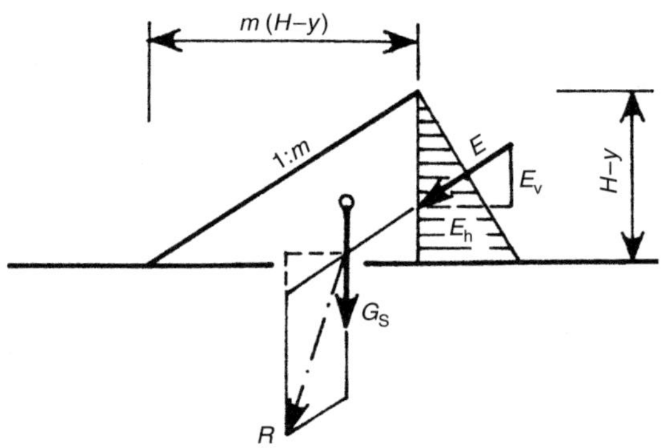

Fig. 38. Forces acting in the shell of an embankment dam[12]

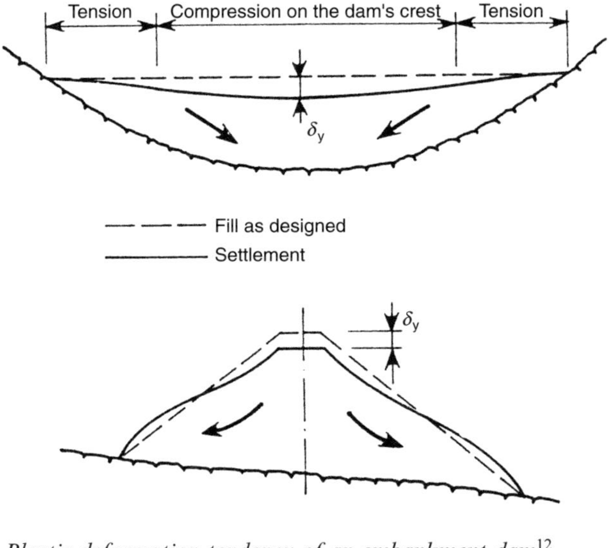

Fig. 39. Plastic deformation tendency of an embankment dam[12]

3.4. Slip safety

The safety of an embankment dam sliding bodily on the subsoil must be distinguished from the slip safety of the dam's shells.

3.4.1. Sliding safety of the dam

The bodily sliding of a dam subjected to water pressure on the near-vertical central core is resisted at the critical height of the existing valley floor (Fig. 40) by the shearing resistance of both the central core and the downstream shell. The average shear stress follows from the effective water pressure according to Equation (18) and the base width B_y of the core and the downstream shell as

$$\tau = \frac{\gamma_W (H - f)^2}{2B_y} \tag{29}$$

With an average overburden pressure of approximately

$$\sigma_z = \gamma H/2 \tag{30}$$

Coulomb's law of friction (neglecting the influence of cohesion ($c' = 0$)) gives the condition of sliding as

$$\frac{\tau}{\sigma_z} = \frac{\gamma_W (H - f)^2}{\gamma H B_y} \leq \tan \phi' \tag{31}$$

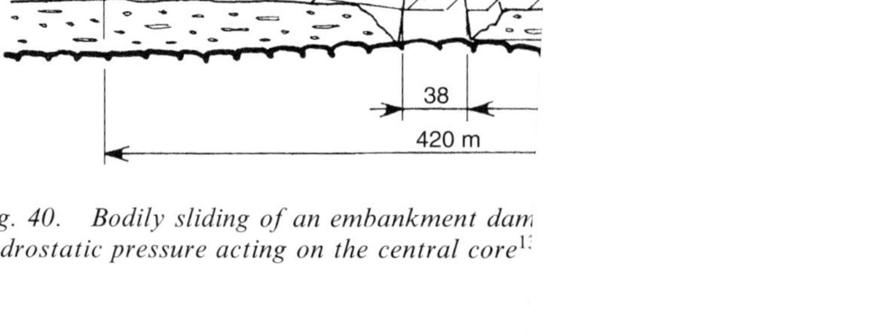

Fig. 40. Bodily sliding of an embankment dam hydrostatic pressure acting on the central core[1:]

3.4.2. Slip safety of shell slopes

In embankment dams with a vertical central core and upstream and downstream shells of the same slope the slip safety of the upstream shell subjected to a rapid lowering of the reservoir level is always smaller than that of the downstream shell because the flowing force of the moving water favours the slip. For this reason the upstream slope is usually flatter than the downstream slope.

In Fig. 41 the deadweight of a slice of the slip body is approximately given by

$$G_D = (\gamma - \gamma_w)\left(\frac{a_1}{2} + \frac{2}{3}a_2\right)b \qquad (32)$$

and the average slope α of the slip circle yields load components of

$$N_D = G_D \cos \alpha \qquad (33a)$$

$$T_D = G_D \sin \alpha \qquad (33b)$$

For the average shear strength $\tan \phi'$ of the slip circle the slip safety of the upstream shell amounts finally to

$$s = \frac{N_D}{T_D} \tan \phi' \qquad (34)$$

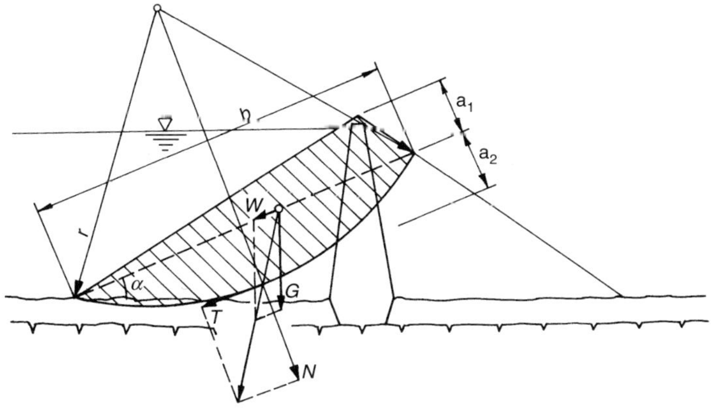

Fig. 41. Slope slip due to rapid lowering of the reservoir level of an embankment dam with a central core

In the case of an earthquake characterized by the horizontal and vertical response accelerations a_{Eh} and a_{Ev}, the force components

$$E_h = \frac{a_{\mathrm{Eh}}}{g} G_{\mathrm{D}} \tag{35a}$$

$$E_{\mathrm{v}} = \frac{a_{\mathrm{Ev}}}{g} G_{\mathrm{D}} \tag{35b}$$

must be added. As will be shown in Section 3.5 the response acceleration a_{E} of the dam will only exceptionally coincide with the ground acceleration.[111]

3.5. Earthquakes

3.5.1. Fundamental period of the dam

The fundamental period of the dam is given by the well known approximation for a single-mass oscillator:[112]

$$T_1 = 1{\cdot}79\sqrt{w_1} \tag{36}$$

The crest displacement w_1 (in metres) is computed for the horizontally acting deadweight of the dam according to Equation (16).

3.5.2. Improved equivalent static load

The relevant parameters of the *seismic coefficient* are considered with four multipliers.[112] With $T_{\mathrm{b}} = 0{\cdot}5$ s for rock, $T_{\mathrm{b}} = 1{\cdot}0$ s for diluvium and $T_{\mathrm{b}} = 2{\cdot}5$ s for alluvium, the first multiplier accounting for the dam's fundamental period is

$$M_1 = T_{\mathrm{b}}/T_1 \leq 1 \tag{37}$$

The dynamic deformation modulus E_{dyn} of the dam's foundation material gives the second multiplier as

$$M_2 = \left(\frac{2}{\log E_{\mathrm{dyn}}}\right)^2 \tag{38}$$

With the ductility of a structure defined as the ratio F of its deformation at rupture to that at the limit of proportionality (for concrete dams $F = 3$, rockfill dams $F = 6$ and earth dams $F = 10$), the third multiplier is given by

$$M_3 = 3/F \tag{39}$$

Due to the varying frequency of a single earthquake the possibility of resonance exists for any earthquake of sufficient duration. If the dam's internal damping is described with the proportion D of the critical damping as in Table 1, the multiplier of a single degree of freedom oscillator excited by an earthquake cannot surpass the theoretical limit of

$$M_4 = 1/2D \tag{40}$$

In reality, possible resonance is always disturbed by the varying frequency of the earthquake. This is taken into account by replacing the ratio f_E/f_1 (f_E is the exciting frequency of the earthquake and f_1 is the natural period of the dam) by the approximation $f_E/f_1 \pm 0{\cdot}05$. Neglecting the only marginal influence of interior damping in the case of resonance, the multiplier cannot surpass the upper bound

$$M_4 = 1/[1 - (f_E/f_1 - 0{\cdot}05)^2] = 10 \tag{41}$$

If b_E denotes the ground acceleration of an earthquake and a_E the response acceleration of the dam, their reciprocal ratio equals the product of the four relevant multipliers

$$a_E/b_E = M_1 M_2 M_3 M_4 \tag{42}$$

Its magnitude for embankment dams lies in the range $0{\cdot}5 < a_E/b_E < 1{\cdot}0$.

Table 1. Internal damping as a proportion D *of critical damping*

	D	
Type of dam	elastic	inelastic
Concrete	0·03	0·10
Rockfill	0·06	0·20
Earth	0·10	0·30

Finally, the seismic coefficient equals the ratio of the dam's dynamic inertial force to the dam's deadweight:

$$C_s = \frac{a_E}{g} = \frac{a_E}{b_E}\frac{b_E}{g} = \frac{E}{G} \tag{43}$$

The ground acceleration can act in any direction between horizontal and vertical (Fig. 42).

3.6. Embankment dams with impervious surfaces

In the case of an embankment dam with a central core the water pressure acts on the core inside the dam. In the case of an embankment dam with a waterproof surface the water pressure acts on the upstream face of the dam which carries the waterproofing medium.

3.6.1. Water pressure

The vertical component of water pressure (Fig. 43)

$$W_v = \gamma_w \frac{Tb_w}{2} = m\gamma_w \frac{T^2}{2} \tag{44}$$

increases the vertical surface settlement by

$$\Delta u_2 = \frac{\gamma_w th}{E_2} \tag{45}$$

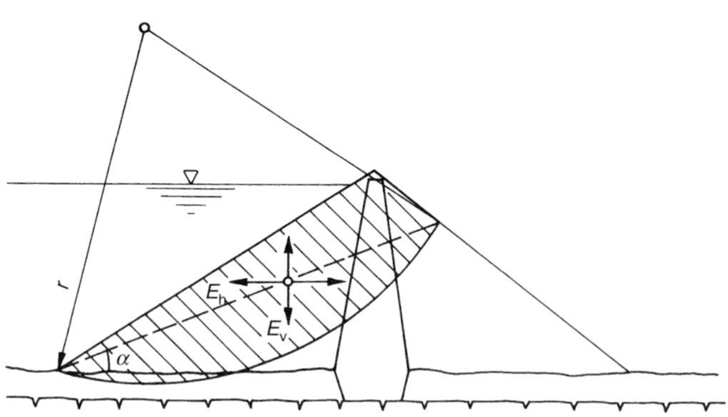

Fig. 42. Horizontal and vertical components of earthquake load acting in a slope slip

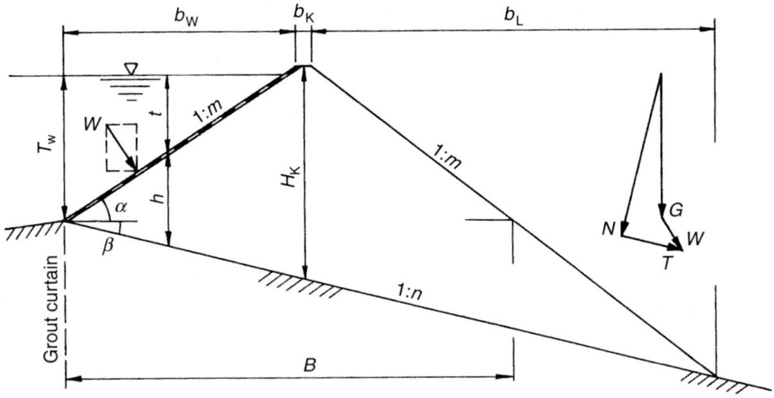

Fig. 43. Cross-section of an embankment dam upon an inclined bedrock with waterproof upstream slope

The horizontal component of water pressure

$$W_h = \gamma_w \frac{T^2}{2} \tag{46}$$

causes, in a horizontal section of the dam, the shear stress

$$\tau = W_h / B \tag{47}$$

3.6.2. Shear safety

The shear safety of a horizontal dam section runs in analogy with Equation (31)

$$\frac{\tau}{\sigma_z} = \frac{\gamma_w T}{\gamma B} \leq \tan \phi' \tag{48}$$

3.6.3. Sliding safety

Finally, the sliding safety of the entire dam body subjected to deadweight and water pressure upon the eventually sloped bedrock (Fig. 43) must be proven

$$T/N < \tan \phi' \tag{49}$$

3.7. Spatial action in narrow valleys

In narrow valleys a three-dimensional load-carrying action[19,20] can be predicted not only with finite elements[16,17] but also with the less expensive grid method.[113]

3.8. Numerical example: Gepatsch Dam (Fig. 44)

The 153 m high Gepatsch Dam in the Kauner Valley, Tyrol, Austria, was built in 1961 – 1964.[107] From 1964 to 1966 the reservoir was filled by one third each year. The high porosity of $n = 29\%$ for both the upstream and downstream shells consisting of roller-compacted quarry material ($\gamma = 19{\cdot}0$ kN/m^3 and $\phi' = 43°$) proved lucky in retrospect. It reduced the silo action[104] in the vertical central core — consisting of landslip material (upstream side enriched with bentonite: $\gamma = 23{\cdot}4$ to $23{\cdot}6$ kN/m^3; $\phi' = 35°$; $n = 17{\cdot}4$ to $18{\cdot}3\%$; $k = 5\times10^{-9}$ to 5×10^{-10} m/s) — due to the large shell settlements (Fig. 45) to such a low value that no horizontal cracks in the core were detected. The perfect performance of the dam's core is confirmed by a very small seepage (in 1966 18 l/s and in 1984 11 l/s).

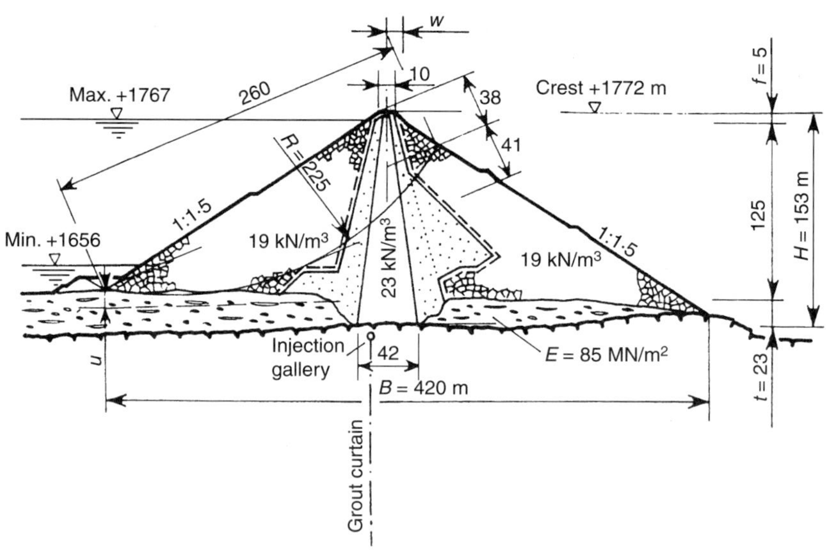

Fig. 44. Cross-section of Gepatsch rockfill dam with zoning[107]

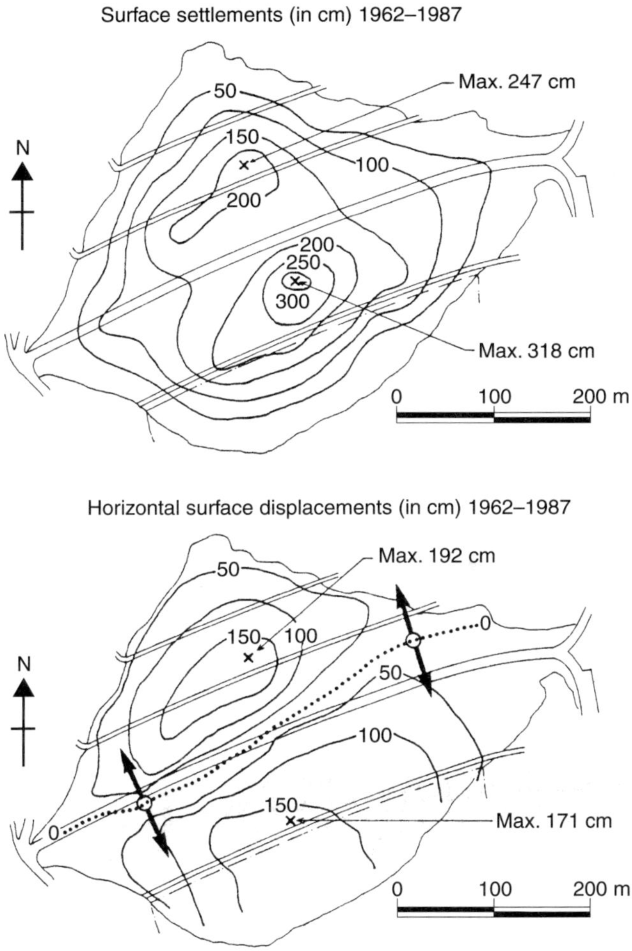

Fig. 45. Measured surface settlements and displacements of Gepatsch Dam up to 1987[114]

3.8.1. Vertical settlements

With the deformation moduli for the different dam zones as defined by the dam's designer in 1967:[10]

- shells $E = 13$ MN/m^2
- core $E = 52$ MN/m^2
- filters $E = 240$ MN/m^2
- valley floor (subsoil) $E = 85$ MN/m^2

the theoretical consolidation settlements (measured values in parentheses) of the dam's surface according to Equation (3) are as follows.

(a) Downstream shell (El. 1719 m)

$$u_{2d} = \frac{0{\cdot}024 \times 23 + 0{\cdot}019 \times 77}{23 + 77} \frac{100^2}{2 \times 13} = 7{\cdot}75\,\text{m}\,(2{\cdot}47\,\text{m})$$

(b) Upstream shell (El. 1715 m)

$$u_{2u} = \frac{0{\cdot}024 \times 26 + 0{\cdot}019 \times 70}{26 + 70} \frac{96^2}{2 \times 13} = 7{\cdot}21\,\text{m}\,(3{\cdot}18\,\text{m})$$

(c) Central core (El. 1772 m)

$$u_{2c} = \frac{0{\cdot}0235 \times 153^2}{2 \times 52} = 5{\cdot}29\,\text{m}\,(1{\cdot}75\,\text{m})$$

The measured values (in parentheses above) up to 1987 (observation period of 23 years) are much smaller. From the latter the true deformation moduli can be deduced as follows.

(a) Downstream shell

$$E_{2d} = 13 \times \frac{7{\cdot}75}{2{\cdot}47} = 41\ \text{MN/m}^2$$

(b) Upstream shell

$$E_{2u} = 13 \times \frac{7{\cdot}21}{3{\cdot}18} = 29\ \text{MN/m}^2$$

(c) Central core

$$E_{2c} = 52 \times \frac{5{\cdot}29}{1{\cdot}75} = 157\ \text{MN/m}^2$$

Long-term measurements[11] for dam section II with the help of Equations (2) and (3) yield the deformation moduli of the construction phase E_1 and of the consolidation phase E_2 as in Table 2.

It follows from the figures in Table 2 that the deformation moduli of the upstream and downstream shells are of the same magnitude, and that the quarried material of the shells was hardly compactable.

Table 2. Deformation moduli for Gepatsch Dam

Shell	H: m	u_1: m	u_2: m	E_1: MN/m^2	E_2: MN/m^2
Upstream	121	2·23	3·18	32	45
Downstream	94	1·72	1·97	25	44

3.8.2. Horizontal displacements and displacement vectors

The measured horizontal displacements (Fig. 45) and vertical settlements (Ref. 114, p. 186, Fig. 3) are listed in Table 3.

The data in Table 3 yield an average ratio of $w/u = 0{\cdot}66$. The plastic deformation tendency of the dam (Fig. 39) can be guessed approximately (Fig. 46) as

$$\tan \alpha' = \frac{B}{6} \times \frac{3}{2H} = \frac{B}{4H} = \frac{2mH}{4H} = \frac{m}{2} = \frac{1{\cdot}5}{2} = 0{\cdot}75$$

Both values (compare with Equation (28))

$$w/u = \tan \alpha'$$

are at least of the same magnitude.

Table 3. Measured horizontal displacements and vertical settlements of Gepatsch Dam

Shell	w: m	u: m	w/u
Upstream	1·71	3·18	0·54
Downstream	1·92	2·47	0·78

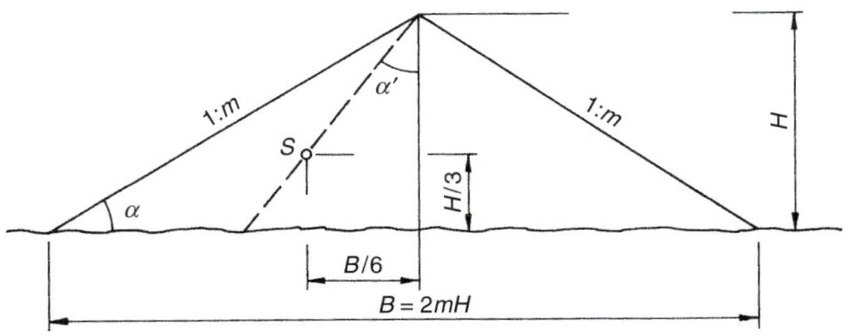

Fig. 46. Average inclination of the deformation vector of half the dam

3.8.3. Silo action of the dam's core

The smaller settlements of the core as compared with those of the shells (see Section 3.8.1) show that the more deformable landslip material of the core is propped by the stiffer filter zones — similar to the fill in a high silo by wall friction. With a core width of $b_c = 42$ m, horizontal earth pressure at rest coefficient $\lambda_0 = 1 - \sin\phi' = 0{\cdot}43$ and wall friction coefficient $\tan\delta = \tan\phi' = 0{\cdot}70$, the vertical bottom pressure of the dam's core according to Equation (4) amounts to

$$p_v = \frac{0{\cdot}024 \times 42}{2 \times 0{\cdot}43 \times 0{\cdot}70} = 1{\cdot}67\ \text{MN/m}^2$$

This value is smaller than the theoretical overburden pressure (Equation (6)) of

$$p_g = 0{\cdot}024 \times 153 = 3{\cdot}67\ \text{MN/m}^2$$

Measured values of the bottom pressures amounted to 1·4 MN/m^2 in the core (Ref. 10, p. 694, Fig. 12) and to 3·9 MN/m^2 in the filter zones (Ref. 11, p. 175, Fig. 3).

3.8.4. Crest displacement for a full reservoir

When considering the dam's deformation due to filling of the reservoir three actions must be distinguished. First, the water pressure on the vertical central core with the average deformation modulus of both core and downstream shell

$$E_m = \frac{0{\cdot}4 \times 41 + 0{\cdot}2 \times 157}{0{\cdot}4 + 0{\cdot}2} = 80\ \text{MN/m}^2$$

and the shear modulus derived from it

$$G_m = 0{\cdot}4E_m = 0{\cdot}4 \times 80 = 32\ \text{MN/m}^2 \tag{50}$$

renders a positive crest displacement in the downstream direction (Equation (19)) of

$$w_1 = \frac{0{\cdot}01(130 - 5)^2}{4 \times 32} \times \frac{130}{420} = 0{\cdot}38\ \text{m}$$

Second, as the submerged upstream shell is subjected to the uplift of the reservoir water the upstream earth pressure is smaller than the downstream one. This difference causes a negative crest displacement in the upstream direction (see Section 3.8.1, $E_{2u} = 29\ \mathrm{MN/m^2}$). This, according to Equation (21), is

$$w_2 = -\frac{0{\cdot}32(0{\cdot}019 - 0{\cdot}010) \times 130^3}{4 \times 0{\cdot}4 \times 29 \times 420} = -0{\cdot}32\ \mathrm{m}$$

Third, compression of the subsoil above the bedrock (eye gneiss) due to the reservoir load (Fig. 5) causes another negative crest displacement. According to Equations (23) and (24) this is

$$w_3 = -0{\cdot}01(130 - 5) \times \frac{23}{85} \times \frac{130}{420} = -0{\cdot}11\ \mathrm{m}$$

The sum of these theoretical crest displacements for a full reservoir, $w = 0{\cdot}38 - 0{\cdot}32 - 0{\cdot}11 = -0{\cdot}05$ m, coincides astonishingly well with the value measured in 1977[115] (Fig. 47) of

$$w_0 = -(0{\cdot}60 + 0{\cdot}11)/2 = -0{\cdot}355\ \mathrm{m}$$

minus the plastic crest displacement of

$$\Delta w = 0{\cdot}55/2 = 0{\cdot}275\ \mathrm{m}$$

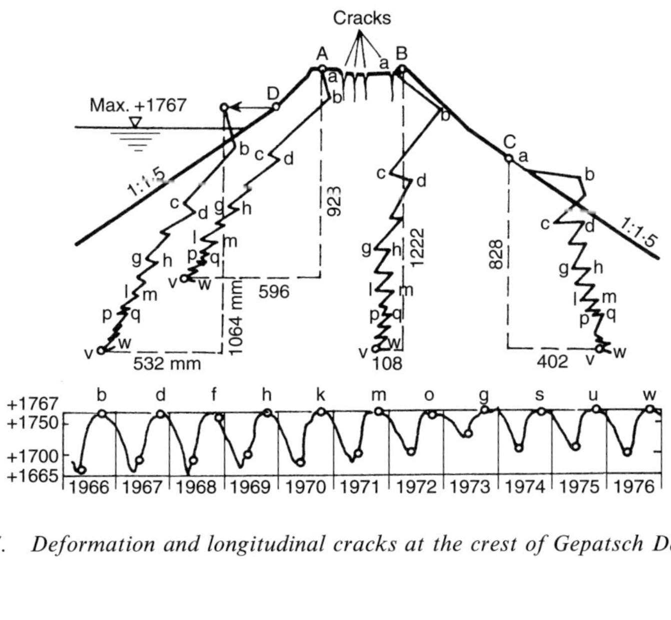

Fig. 47. Deformation and longitudinal cracks at the crest of Gepatsch Dam[115]

or, in total,

$$w = -0{\cdot}355 + 0{\cdot}275 = -0{\cdot}08 \text{ m}$$

3.8.5. Shear safety of the downstream shell

With a critical depth (Equation (5)) of

$$z_0 = \frac{1{\cdot}67}{0{\cdot}024} = 69{\cdot}6 \text{ m}$$

and with consideration of silo action after Equation (7), the earth pressure of the central core amounts to

$$E_0 = \frac{0{\cdot}024}{2}[130^2 - (130 - 69{\cdot}6)^2] \times 0{\cdot}43 = 68{\cdot}4 \text{ MN/m}$$

Together with the horizontal water pressure according to Equation (8)

$$W = \frac{0{\cdot}01}{2}(130 - 5)^2 = 78{\cdot}1 \text{ MN/m}$$

the average shear stress at the elevation of the valley floor (Equation (9)) is

$$\tau = \frac{68{\cdot}4 + 78{\cdot}1}{204} = 0{\cdot}72 \text{ MN/m}^2$$

The average vertical stress of the downstream shell due to its deadweight (Equation (10))

$$\sigma_z = 0{\cdot}0197 \times 68 = 1{\cdot}34 \text{ MN/m}^2$$

yields, according to Equation (12), the shear condition

$$\frac{\tau}{\sigma_z} = \frac{0{\cdot}72}{1{\cdot}34} = 0{\cdot}537 < \tan 43^\circ = 0{\cdot}933$$

The existing shear safety along a horizontal shear plane is then sufficient, i.e.

$$s = \frac{0{\cdot}933}{0{\cdot}537} = 1{\cdot}74$$

3.8.6. *Slip safety of the upstream slope*

The slip circle in Fig. 44 approximates the deadload of the 1 m thick slip body (after Equation (32)) as

$$G_D = (0{\cdot}0197 - 0{\cdot}01)\left(\frac{38}{2} + \frac{2}{3} \times 41\right) \times 260 = 116{\cdot}8\ \text{MN/m}$$

According to Equations (33), an average slope angle of the slip circle $\alpha = 25^\circ$ yields the deadload components

$$N_D = 116{\cdot}8 \times 0{\cdot}906 = 105{\cdot}9\ \text{MN/m}$$
$$T_D = 116{\cdot}8 \times 0{\cdot}423 = 49{\cdot}4\ \text{MN/m}$$

The average shear resistance of the slip circle $\tan\phi' = \tan 39^\circ\ 20' = 0{\cdot}82$ finally gives the slip safety of the dam's upstream slope (Equation (34)) as

$$s = \frac{105{\cdot}9}{49{\cdot}4} \times 0{\cdot}82 = 1{\cdot}76$$

Because of the shell's high porosity ($n = 29\%$) the usually decisive case of rapid lowering of the reservoir level is not relevant for the Gepatsch Dam.

3.8.7. *Influence of consolidation on the characteristic values of dam material*

Consolidation of the dam after the construction phase generally causes a small change in densities and friction angles, as shown in Table 4.

Table 4. Effects of dam consolidation after construction

	$T = 0$[107]		$T = \infty$[11]	
Dam zone	γ_0: kN/m^3	ϕ_0: degrees	γ_∞: kN/m^3	ϕ_∞: degrees
Central core	24·0	29	24·1	35
Filters	24·0	45	24·6	40
Shells	16·9	45	19·7	43
Subsoil	23·4	36	23·4	36

3.9. Numerical example: Oroville Dam

The Oroville Dam[116–119] on the Feather River in the hills of the western slope of the Sierra Nevada in California (about 140 km north of Sacramento) is the highest rockfill dam in the USA. Although the possibility of earthquakes with a ground acceleration of $b_E = 0{\cdot}1g$ had been considered in the design, the occurrence of several earthquakes of this magnitude in the summer of 1975 came as a true surprise. As a consequence, excavation for the Auburn Dam — intended to become the largest arch dam in the world — which was already in progress was ceased.[120]

3.9.1. Earthquakes in summer 1975

Apart from different fore- and after-shocks, three main-shocks occurred on 1 August, 5 August and 27 September[121] as detailed in Table 5.

On 1 August, although the Oroville Dam was well equipped with four accelerometers of the Californian Department of Water Resources and three of the US Geologic Survey, six piezometers and five groups of stress meters, the electricity supply of these instruments was overcharged as a result of their automatic switch on. Thus, data for the most interesting five to six seconds at the beginning of the earthquake — time in which the auxiliary electricity supply was turned on — are missing. On 5 August it was again not possible to record all the data. Only on 27 September three of the four accelerometers of the Californian Department of Water Resources succeeded in recording a complete set of data (the three accelerometers of the US Geologic Survey were again out of action). The data are given in Table 6.

The horizontal displacement of the dam's crest in the transverse direction was measured to be 15 mm and the vertical crest settlement to be 10 mm on 1 August 1975. The natural period of vibration

Table 5. Scale of main-shocks of earthquakes in California in 1975

Earthquake date	Richter magnitude	Distances: km	Depth: km
1 August	5·7	11·1	8·8
5 August	4·7	6·8	8·9
27 September	4·6	3·5	5·6

Table 6. Data recorded on 27 September 1975

		a_{Eh}/g	
Height: m a.s.l.	Location	1 August	27 September
281	Crest	(0·24)	0·12
244	—	Out of service	Out of service
207	—	Out of service	Out of service
46	Bedrock	0·08	0·04

was $T_1 = 0{\cdot}8$ s. These few results demonstrate the difficulty of measuring earthquake data.

3.9.2. Approximate analysis

If the real cross-section of the dam is approximated by a wide wedge of identical base width and height (Fig. 48) and the measured (or estimated) response accelerations of the Oroville Dam: are used (Table 7), the horizontal crest displacement can be predicted from the closed form of Equation (15) as

$$\max\ w^{E_h} = 0{\cdot}13 \times \frac{0{\cdot}0235 \times 219^2}{4 \times 1970} = 0{\cdot}0186 \text{ m}$$

In this calculation the average density of the dam was introduced as $\gamma = 23{\cdot}5$ kN/m^3 and the dynamic shear modulus with the approximate value[121] of

$$G_{dyn} = 9H = 9 \times 219 = 1970 \text{ MN/m}^2 \tag{51}$$

was used. The vertical crest settlement, for a dynamic deformation modulus of $E_{dyn} = 1970/0{\cdot}4 = 4930$ MN/m^2, is predicted by Equation

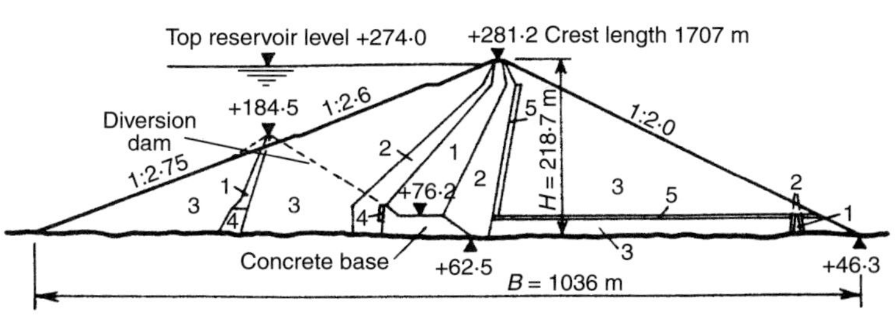

Fig. 48. Cross-section of Oroville Dam with zoning[124]

Table 7. Measured (estimated) response accelerations of the Oroville Dam

	a_E/g	
Location	Horizontal	Vertical
Crest	0·24	(0·16)
Centre of gravity	(0·13)	(0·09)
Bedrock	0·08	(0·05)

(3) as

$$\max\ u_2^{E_v} = 0{\cdot}09 \times \frac{0{\cdot}0235 \times 219^2}{2 \times 4930} = 0{\cdot}0103\ \text{m}$$

With the help of the crest displacement due to the dam's deadload, assumed to act in the horizontal direction

$$u_1 = \frac{\max\ w^{E_h}}{a_{E_h}/g} = \frac{0{\cdot}0186}{0{\cdot}13} = 0{\cdot}143\ \text{m} \qquad (52)$$

the natural period of vibration after Equation (36) amounts to $T_1 = 1{\cdot}79\sqrt{0{\cdot}143} = 0{\cdot}68$ s.

A comparison of measured and predicted values of displacement, settlement and natural period of the Oroville Dam during the earthquake on 1 August 1975 is given in Table 8.

Table 8 shows that there is no sense — at least for preliminary analyses and checks — in using the certainly more up-to-date but a thousand times more expensive method of finite elements[122] instead of the very simple approximation with closed formulae.

3.10. Numerical example: Van Norman Dams

During the San Fernando earthquake of 9 February 1971[82] the embankment dams of the Upper and Lower Van Norman Reservoirs

Table 8. Comparison of measured and predicted parameters of Oroville Dam

		Prediction	
	Measurement	Finite element	Approximation
max w: mm	15	15	18·6
max u_2: mm	10	15	10·3
T_1: s	0·8	0·75	0·68

of the water supply of Los Angeles in California were exposed to horizontal ground accelerations of $b_{Eh}/g = 0{\cdot}5$ to $0{\cdot}6$.[123] As both dams were damaged (the lower one considerably,[99]) this event serves well for a comparison of different methods of analysis.

3.10.1. Improved equivalent static load

Using the crest displacement due to the deadload — assumed to act in the horizontal direction — (Equation (16)) and the dynamic shear modulus (Equation (51)) yields the following.

(a) Upper Van Norman Dam

$$w_1 = \frac{0{\cdot}019 \times 22{\cdot}39^2}{4 \times 9 \times 22{\cdot}39} = 0{\cdot}0118 \text{ m}$$

(b) Lower Van Norman Dam

$$w_1 = \frac{0{\cdot}019 \times 39{\cdot}80^2}{4 \times 9 \times 39{\cdot}80} = 0{\cdot}0210 \text{ m}$$

The natural period of vibration is given by Equation (36):

(a) Upper Van Norman Dam

$$T_1 = 1{\cdot}79\sqrt{0{\cdot}0118} = 0{\cdot}194 \text{ s}$$

(b) Lower Van Norman Dam

$$T_1 = 1{\cdot}79\sqrt{0{\cdot}0210} = 0{\cdot}259 \text{ s}$$

As the underground of both dams consists of bedrock covered by alluvium of different thickness, the first multiplier (Equation (37)) for both dams is $M_1 = 1{\cdot}0$.

With the dynamic deformation modulus of the ground $E_{dyn} = 200$ MN/m^2, the relevant multiplier for both dams is given by Equation (38) as

$$M_2 = \left(\frac{2}{2{\cdot}301}\right)^2 = 0{\cdot}76$$

The ductility $F = 10$ of both dams which were partly hydraulic fill and partly roller-compacted yields the multiplier (Equation (39))

$$M_3 = 3/10 = 0{\cdot}30$$

The ratio of internal damping to the so-called critical damping of $D = 0{\cdot}15$ gives the multiplier (Equation (40))

$$M_4 = \frac{1}{2 \times 0{\cdot}15} = 3{\cdot}33 < 10$$

Then, Equation (42) gives the ratio of the response acceleration of both dams to the ground acceleration as

$$a_E/b_E = 1{\cdot}0 \times 0{\cdot}76 \times 0{\cdot}30 \times 3{\cdot}33 = 0{\cdot}76$$

For an average ground acceleration of $b_{Eh}/g = 0{\cdot}55$ the seismic coefficient after Equation (43) is

$$C_s = 0{\cdot}76 \times 0{\cdot}55 = 0{\cdot}42$$

3.10.2. Slope slip

The submerged deadload of the slip body on the critical upstream side of the dam (Fig. 41) for the

(a) Upper Van Norman Dam

$$G_D = (0{\cdot}019 - 0{\cdot}010) \times 68{\cdot}5 \times \left(\frac{10{\cdot}0}{2} + \frac{2}{3} \times 8{\cdot}3\right) = 6{\cdot}49 \text{ MN/m}$$

(b) Lower Van Norman Dam

$$G_D = (0{\cdot}019 - 0{\cdot}010) \times 110{\cdot}0 \times \left(\frac{15{\cdot}6}{2} + \frac{2}{3} \times 16{\cdot}6\right) = 18{\cdot}68 \text{ MN/m}$$

must be combined (Figs 49 and 50) with the equivalent static load of the earthquake (here assumed to act in the horizontal direction only Fig. 42). This gives the following

(a) Upper Van Norman Dam

$$E_h = 0{\cdot}42 \times 6{\cdot}49 = 2{\cdot}73 \text{ MN/m}$$

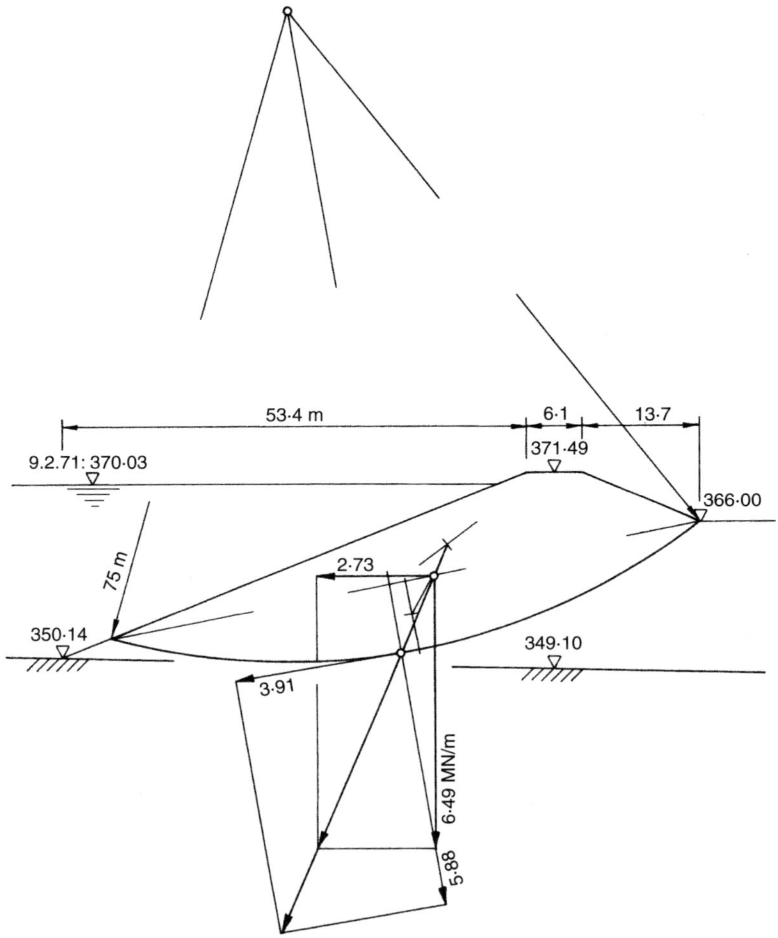

Fig. 49. Slope slip at the Upper Van Norman Reservoir with acting forces

(b) Lower Van Norman Dam

$$E_h = 0{\cdot}42 \times 18{\cdot}68 = 7{\cdot}85 \text{ MN/m}$$

The normal and tangential components N and T of the resulting force R give the slope of the latter at the slip circle as follows.

(a) Upper Van Norman Dam

$$T/N = 3{\cdot}91/5{\cdot}88 = 0{\cdot}665$$

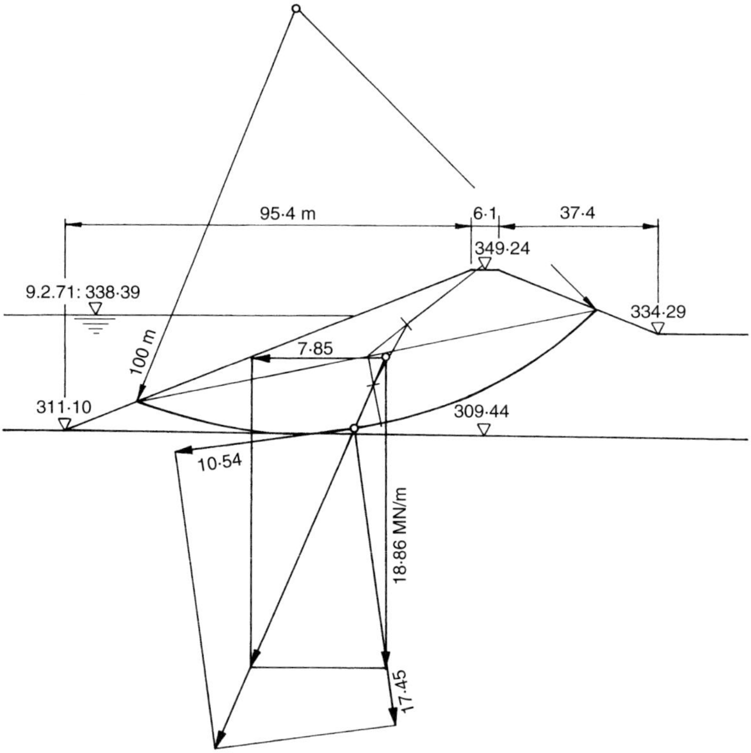

Fig. 50. Slope slip at the Lower Van Norman Reservoir with acting forces

(b) Lower Van Norman Dam

$$T/N = 10{\cdot}54/17{\cdot}45 = 0{\cdot}604$$

At the Upper Van Norman Dam a crest displacement of 1·5 m and a crest settlement of 0·9 m were measured. It was not possible to prove a slip by mere surface inspection. As the three piezometers showed pore pressures in excess of the reservoir level (Ref. 99, p. 661, Fig. 9) liquefaction of this dam can be suspected. At the Lower Van Norman Dam the complete agreement between the real breaking edge of the slip and the predicted one is impressive.

3.11. Numerical example: Mica Dam (Fig. 51)

The Mica Dam was built in four fill campaigns from 1969 to 1972 in a narrow gorge of the Columbia River in British Columbia, Canada,

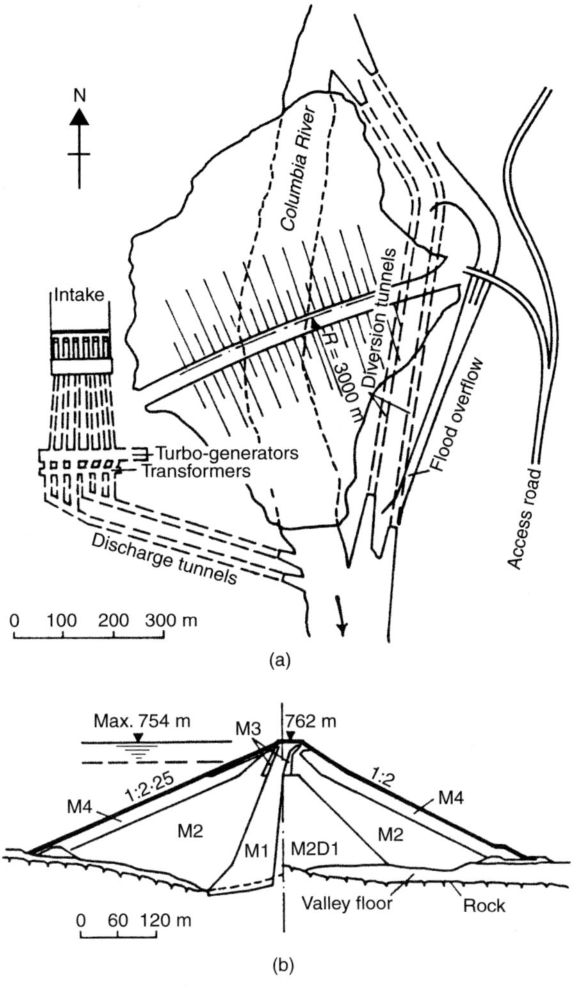

Fig. 51. The Mica Dam on the Columbia River in Canada: (a) *plan and* (b) *main cross-section*[9]

about 130 km to the north of Revelstoke.[106,125,126] It consists of a near-vertical core of glacial till as well as upstream and downstream shells of sand and gravel. Because of its size (dam volume 33×10^6 m^3) and importance (the Mica Dam was, at the time of its completion, at 244 m the highest embankment dam in the world) a large number of measuring instruments was installed. 88 surface bolts, 15 vertical and 8 horizontal displacement meters, 206 stress gauges, 53 pressure meters, 67 piezometers, three strong-motion and six peak accelerometers were expected to yield information on the true behaviour of this dam.

3.11.1. Construction and consolidation settlements

With the assumption that the core settlement is not influenced by the 13 times stiffer shells, the biggest construction settlement at midheight would (for $\gamma = 21{\cdot}5$ kN/m^3 and $E = 35$ MN/m^2) amount to (Equation (2))

$$u^0_{1c} = \frac{0{\cdot}0215 \times 244^2}{4 \times 35} = 9{\cdot}14 \text{ m}$$

The vertical displacement meter MV 8 measured only 1·84 m.[106,126] The construction settlement of the shell — not influenced by the adjoining core — would (with $\gamma = 24{\cdot}3$ kN/m^3 and $E = 460$ MN/m^2) amount to

$$u^0_{1s} = \frac{0{\cdot}0243 \times 244^2}{4 \times 460} = 0{\cdot}79 \text{ m}$$

Whereas the vertical displacement meter MV 9 measured the slightly larger value of 0·84 m.

3.11.2. Silo action

From the above measurements it can be deduced that the softer core of glacial till is propped by the stiffer shells (Fig. 52) due to wall friction.

Considering the two-dimensional case only, an average core width of $b_c = 46$ m, the internal angles of friction of $\phi = 34°$ for the core and $\phi = 42°$ for the shells, ratio of horizontal to vertical pressure at rest of $\lambda_0 = 1 - \sin\phi = 0{\cdot}44$ and a coefficient of wall friction between core and shell of $\tan\delta = \tan\phi = 0{\cdot}675$, Equation (4) gives a vertical pressure of

$$p_v = \frac{0{\cdot}0215 \times 46}{2 \times 0{\cdot}44 \times 0{\cdot}675} = 1{\cdot}665 \text{ MN/m}^2$$

whereas the overburden pressure (given by Equation (6) is

$$p_g = 0{\cdot}0215 \times 244 = 5{\cdot}246 \text{ MN/m}^2$$

which is more than three times larger. With the critical depth

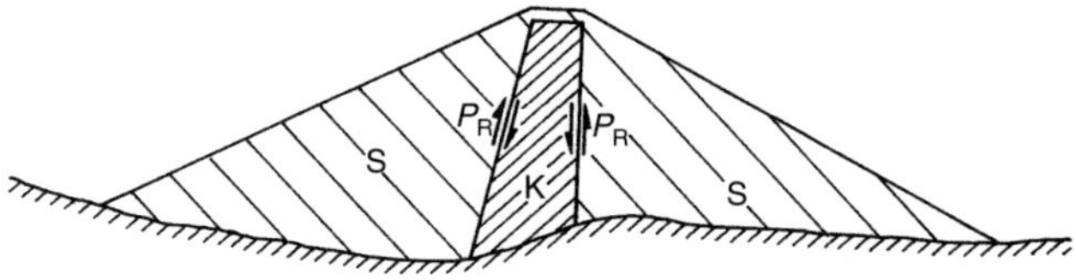

Fig. 52. Silo action of the shells S of a rockfill dam with a clay core K[9]

according to Equation (5) given by

$$z_0 = \frac{1{\cdot}665}{0{\cdot}0215} = 77{\cdot}44 \text{ m}$$

the average vertical pressure in the core is

$$\bar{p}_v = p_v\left(1 - \frac{z_0}{2H}\right) = 1{\cdot}665\left(1 - \frac{77{\cdot}44}{2 \times 244}\right) = 1{\cdot}40 \text{ MN/m}^2 \quad (53)$$

and the construction settlement — with due consideration of silo action — is then

$$u_{1c} = \frac{\bar{p}_v}{p_g} u_{1c}^0 = \frac{1{\cdot}401}{5{\cdot}246} \times 9{\cdot}14 = 2{\cdot}44 \text{ m} \quad (54)$$

In reality the somewhat smaller value of 1·84 m was measured.

Due to wall friction the shell adjoining the core had to carry additionally a portion of the core's weight. Thus, it comes as no surprise that the construction settlement of the shell adjoining the core is bigger (MV 9: $u_{1s} = 0{\cdot}84$ m) than that predicted by approximate analysis disregarding silo action (MV 8: $u_{1s}^0 = 0{\cdot}79$ m).

3.11.3. Vertical tension in the core

The difference between predicted construction settlements without and with silo action yields the following approximate value of vertical tensile stress in the core

$$\sigma_z = \frac{2(u_{1c}^0 - u_{1c})}{H} E_c = \frac{2(9{\cdot}14 - 2{\cdot}44)}{244} \times 35 = 1{\cdot}92 \text{ MN/m}^2 \quad (55)$$

Permeability tests carried out gave such high seepage losses (Ref. 126, p. 240) that core cracks are considered highly probable.

3.11.4. Pore pressure (Fig. 53)

Measured pore pressures in the Mica Dam were so small that they hardly influenced the construction settlements. The consolidation of the core would have provoked vertical settlements in the meantime. Their magnitude can be estimated with the help of the ratio of the pore pressure at foundation level ($p_w = 0{\cdot}69$ MN/m^2) to the overburden

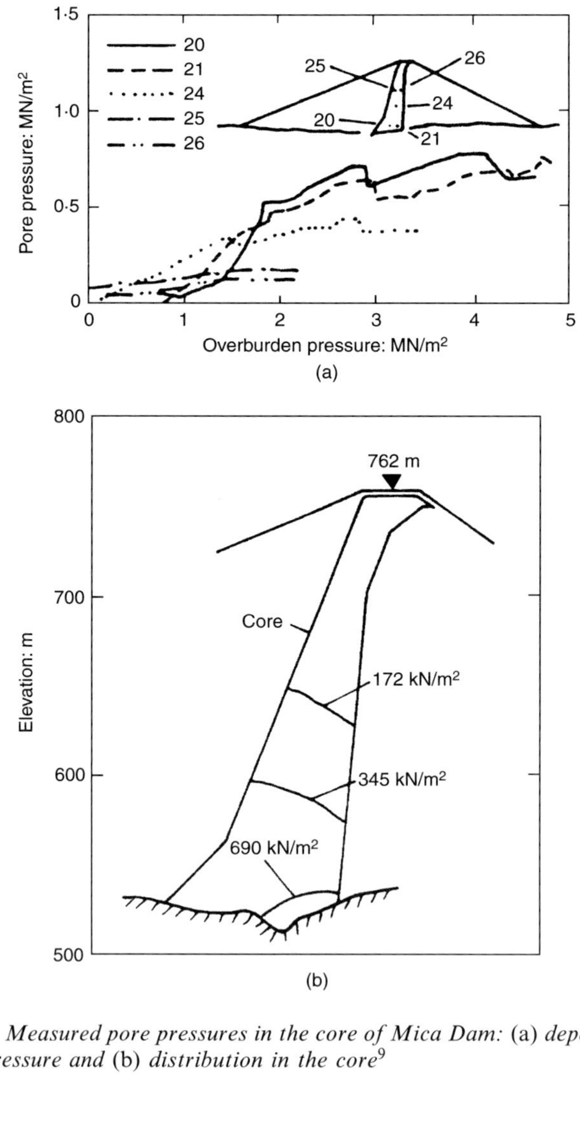

Fig. 53. Measured pore pressures in the core of Mica Dam: (a) *dependence on overburden pressure and* (b) *distribution in the core*[9]

pressure ($p_g = 5{\cdot}246$ MN/m^2) with due consideration of silo action ($u_{1c} = 2{\cdot}44$ m), i.e.

$$\Delta u_{1c} = u_{1c}\frac{p_w}{p_g} = 2{\cdot}44 \times \frac{0{\cdot}69}{5{\cdot}246} = 0{\cdot}32 \text{ m} \tag{56}$$

at midheight.

3.12. Numerical example: Sheffield Dam (Fig. 54)

The Sheffield Dam, at only 7·6 m high, was built for the water supply of Santa Barbara in California in the winter of 1917. It consisted uniformly of silty sand with a silt content (below 0·02 mm) of 40 to 60%. The fill material was dredged from the reservoir. It was compacted only by passing transport vehicles. The waterproofing was warranted by a 1·2 m thick layer of clay on the upstream face. It reached to the bedrock in a 1·8 m deep cut-off trench. The surface of the clay was revetted with a thin concrete slab.

With posterior soil explorations[127] it was found that the only 1·5 to 1·8 m thick subsoil above the bedrock was much looser (dry density 14·4 kN/m^3 in the uppermost 0·5 m than further down (16·2 kN/m^3).

3.12.1. Earthquake of 29 June 1925

As no strong-motion seismographs existed at the time of the earthquake, this earthquake was later given a Richter magnitude of $M = 6{\cdot}3$ on the basis of observations at more distant stations. The duration of the main-shock lasted 15 s. The maximum horizontal ground acceleration was estimated as $b_{Eh}/g = 0{\cdot}15$ and the natural period of vibration as $n_1 = 3$ s^{-1}.[128] The reservoir content of 110×10^3 m^3 was completely emptied. The subsoil was saturated. The earthquake initiated liquefaction of the subsoil on which a length of 110 m of the Sheffield Dam slid about 30 m downstream.[129]

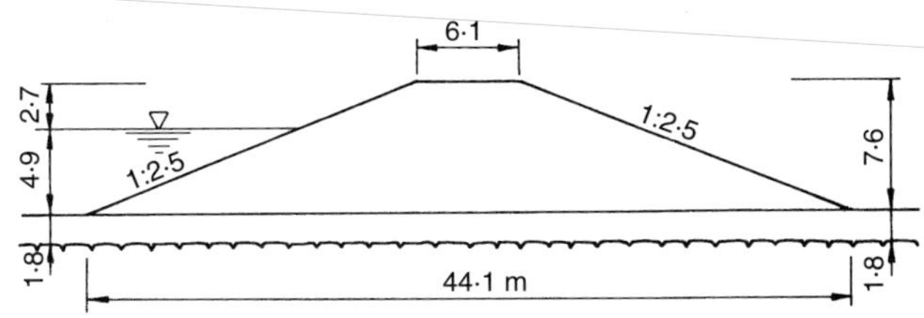

Fig. 54. Cross-section of Sheffield Dam[14]

3.12.2. *Improved equivalent static load*

The fictional crest displacement due to the deadload, assumed to act in the horizontal direction, of (Equation (16))

$$w_1 = \frac{0{\cdot}0172 \times 7{\cdot}6^2}{4 \times 9 \times 7{\cdot}6} = 0{\cdot}00363 \text{ m}$$

yields a natural period of vibration from Equation (36) of

$$T_1 = 1{\cdot}79\sqrt{0{\cdot}00363} = 0{\cdot}108 \text{ s}$$

Then, according to Equation (37), the multiplier of this earth dam upon alluvium becomes

$$M_1 = \frac{2{\cdot}5}{0{\cdot}108} > 1{\cdot}0$$

Equation (50) gives a deformation modulus of

$$E = 9 \times 7{\cdot}6/0{\cdot}4 = 171 \text{ MN/m}^2$$

This yields the second multiplier, after Equation (38), of

$$M_2 = \left(\frac{2}{2{\cdot}233}\right)^2 = 0{\cdot}80$$

With the ductility $F = 10$ of this hardly compacted earth dam Equation (39) gives

$$M_3 = 3/10 = 0{\cdot}30$$

With a damping ratio $D = 0{\cdot}15$, Equation (40) yields

$$M_4 = \frac{1}{2 \times 0{\cdot}15} = 3{\cdot}33 < 10$$

Equation (42) gives the ratio of the response acceleration of the Sheffield Dam to the ground acceleration as

$$a_E/b_E = 1{\cdot}0 \times 0{\cdot}80 \times 0{\cdot}30 \times 3{\cdot}33 = 0{\cdot}80$$

which finally gives the seismic coefficient from Equation (43) as

$$C_s = 0{\cdot}80 \times 0{\cdot}15 = 0{\cdot}12$$

From the deadload of a 1·0 m thick slice of the Sheffield Dam

$$G_D = 17{\cdot}2 \times 7{\cdot}6 \times \frac{6{\cdot}1 + 44{\cdot}1}{2} = 3281 \text{ kN/m}$$

the horizontal component of the improved equivalent static load follows as

$$E_h = 0{\cdot}12 \times 3281 = 394 \text{ kN/m}$$

3.12.3. Safety against sliding

The reservoir fill (water depth 4·9 m) yields the horizontal component of the water load (hydrostatic pressure) as

$$W_h = 10 \times \frac{4{\cdot}9^2}{2} = 120 \text{ kN/m}$$

and the vertical component on the upstream slope 1 in 2·5 as

$$W_v = 120 \times 2{\cdot}5 = 300 \text{ kN/m}$$

The average shear stress at the dam base

$$\tau = \frac{394 + 120}{44{\cdot}1} = 11{\cdot}7 \text{ kN/m}^2$$

is smaller than the existing shear strength of the saturated subsoil without liquefaction ($\tan \phi' = 0{\cdot}231$ and $c' = 10 \text{ kN/m}^2$) which is, according to Equation (11)

$$\tau_u = \frac{3281 + 394}{44{\cdot}1} \times 0{\cdot}231 + 10 = 29{\cdot}3 \text{ kN/m}^2$$

but bigger than the shear strength of the liquefied subsoil of $\tau_u = 0$.

3.13. Numerical example: Oschenik Dam (Fig. 55)

The Oschenik rockfill dam on the south-western slope of the Sonnblick Massif in the High Tauern of Carinthia, Austria was constructed in four stages in the years 1971 to 1979. Because of the dam's altitude of 2300 m above sea level (a.s.l.) the annual construction period lasted only about 100 days. The upstream membrane of asphaltic concrete permitted filling of the reservoir after the first construction period in the autumn of 1972. The dam is founded

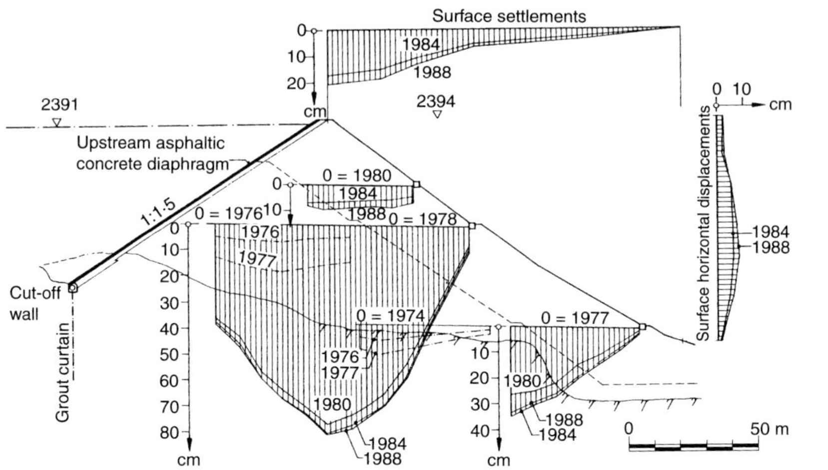

Fig. 55. Cross-section of Oschenik Dam with measured settlements and horizontal displacements[114]

on a natural threshold of granitic gneiss which forms the embankment of a natural lake.

3.13.1. The rockfill dam

The bedrock below the first two stages of the dam construction was covered with moraine material up to a depth of 12 m. The fill material of these first two stages consisted in the beginning of moraine material and later of quarried granitic-gneiss. In the third stage (1974 to 1976) with a fill volume of 1×10^6 m^3, quarried rock of high strength was used. For the final (fourth) construction stage the Little Oschenik Lake had to be emptied first. The rockfill, in layers of 1·3 m thickness with a maximum particle size of 1·0 metre, was compacted with vibratory rollers of 45 tonnes. The characteristic values of the rockfill are given in Table 9.

The bedrock surface was comparatively smooth. In the analysis only a reduced angle of friction of $\phi = 35°$ was used.

Although the vertical grout curtain reaches only 25 m into the compact granitic gneiss, the seepage of only 5 l/s (for a full reservoir) is very low.

3.13.2. Vertical settlements

From the measured construction settlement of $u_1 = 0{\cdot}65$ m at midheight of the 80 m high fill, Equation (2) gives the deformation

Table 9. Characteristic values of rockfill of Oschenik Dam

Fill material	γ: kN/m^3	ϕ: degrees
Moraine	23	38
Quarried rock	21	41 – 45

modulus of the shell as

$$E_1 = \frac{0{\cdot}022 \times 80^2}{4 \times 0{\cdot}65} = 54 \text{ MN/m}^2$$

With the measured consolidation settlement of the dam's crest of $u_2 = 0{\cdot}21$ m, Equation (3) gives

$$E_2 = \frac{0{\cdot}022 \times 80^2}{2 \times 0{\cdot}21} = 335 \text{ MN/m}^2$$

Compared with the Gepatsch Dam (Section 3.8.1), the shells of which consisted entirely of quarried material, the deformation modulus of the construction settlement E_1 of Oschenik Dam is about twice as big and that of the consolidation settlement E_2 an order of magnitude larger. Both observations indicate the much better compaction of the Oschenik Dam.

3.13.3. Horizontal displacements

The static shear modulus (after Equation (50)) of

$$G = 0{\cdot}4 \times 335 = 134 \text{ MN/m}^2$$

yields the following crest displacements of the Oschenik Dam resting on a bedrock with an average downstream slope of 1 in 5·7 (Fig. 56).

(a) Due to deadload, Equation (16) gives

$$\max\ w = \frac{0{\cdot}022 \times 80^2}{4 \times 134} \times \frac{1}{5{\cdot}7} = 0{\cdot}046 \text{ m}$$

(b) Due to full reservoir, Equation (19) gives

$$\max\ w = \frac{0{\cdot}01 \times 60{\cdot}6 \times 77}{4 \times 134} \times \frac{80}{262} = 0{\cdot}026 \text{ m}$$

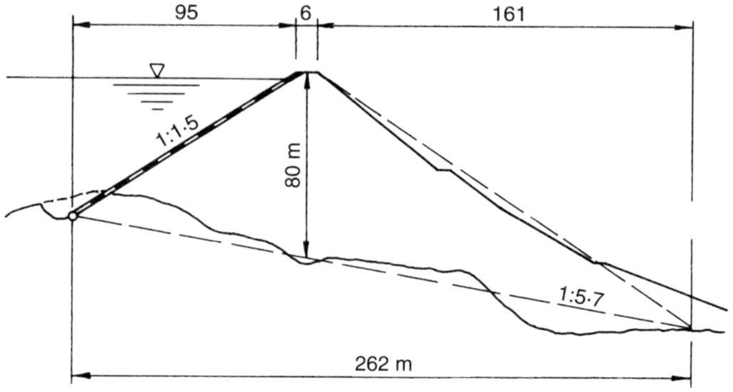

Fig. 56. Simplified cross-section of Oshenik Dam as used in the approximate analysis

In 1988 the following horizontal displacements of the downstream face were observed (Ref. 114, p. 197, Fig. 2):

(a) at midheight, 0·08 m
(b) at the crest, 0·03 m.

3.13.4. Safety against sliding

For the average bedrock slope of $\tan\beta = 1/5{\cdot}7 = 0{\cdot}1754$ and $\beta = 9^\circ\ 57'$, the deadload components of

$$N_g = 0{\cdot}022 \times 80 \times \left(\frac{6+262}{2}\right) \times 0{\cdot}9848 = 232{\cdot}3\ \mathrm{MN/m}$$

$$T_g = 0{\cdot}022 \times 80 \times \left(\frac{6+262}{2}\right) \times 0{\cdot}1737 = 41{\cdot}0\ \mathrm{MN/m}$$

and the waterload components ($\alpha = 33^\circ\ 40'$ and $\alpha + \beta = 43^\circ\ 37'$) of

$$N_w = 0{\cdot}01 \times \left(\frac{60 \times 90}{2}\right) \times 0{\cdot}7240 = 19{\cdot}5\ \mathrm{MN/m}$$

$$T_w = 0{\cdot}01 \times \left(\frac{60 \times 90}{2}\right) \times 0{\cdot}6898 = 18{\cdot}6\ \mathrm{MN/m}$$

yield the shear condition (after Equation (34)) for $c = 0$ as

$$\frac{T}{N} = \frac{41{\cdot}0 + 18{\cdot}6}{232{\cdot}3 + 19{\cdot}5} = 0{\cdot}237 < \tan 35^\circ = 0{\cdot}700$$

The safety against bodily sliding of the dam on the sloped bedrock finally amounts to

$$s = \frac{0{\cdot}700}{0{\cdot}237} = 2{\cdot}95$$

which is by far sufficient.

4. Gravity dams

4.1. Dam types

Principally, *massive* gravity dams must be distinguished from *structured* ones. Because of the required durability of dams, only massive gravity dams were built in the beginning of dam construction, sometimes with curvature in plan (i.e. a convex upstream face). Even before 1586 the 20·7 m high Almanza Dam and from 1579 to 1594 the 41·0 m high Alicante Dam were built in Spain.[130] A recent development is the gravity dam with a base cavern, and four examples of this type exist in Austria.[115]

For economic reasons structured gravity dams appeared at the turn of the last century. Their two main structural elements are the proper *damming deck* and the *buttresses*. The oldest deck type is in the form of multiple arches (1800 Meer Allum Dam in India, height $H = 11{\cdot}9$ m and crest length $L = 800$ m; 1898 Belubula Dam in New South Wales, Australia, $H = 18{\cdot}3$ m and $L = 131$ m; 1917 to 1922 Tirso Dam in Sardinia, Italy, $H = 69{\cdot}5$ m and $L = 283$ m).[91] Early in this century *Ambursen* introduced the reinforced concrete flat slab as deck (1904 Schuylerville Dam in New York, $H = 8{\cdot}5$ m and $L = 75$ m; 1907 Ellsworth Dam in Maine, USA, $H = 21{\cdot}9$ m and $L = 137$ m; 1913 Jordan River Dam near Vancouver in Canada, $H = 38{\cdot}4$ m and $L = 230$ m).[91] The next development of the deck as a thickening of the buttress' upstream edge was due to *Noetzli* (1927 Don Martin Dam in Mexico, $H = 35{\cdot}0$ m with the closed downstream face acting as spillway; 1948 to 1953 Ben Métir Dam in Tunisia, $H = 75{\cdot}0$ m and $L = 495$ m).[131] The most recent development of the hollow buttress dam was initiated by *Stucky* who took up an idea proposed by *Figari* in 1900 (1929 – 1935 Dixence Dam in Switzerland, $H = 87$ m, $L = 420$ m, with variable curvature in plan; 1950 – 1953 Ancipa Dam in Sicily, $H = 111{\cdot}5$ m and $L = 253$ m).[131]

4.2. Massive gravity dams

The cross-section of massive gravity dams developed from the trapezoid of the Puentes Dam in Spain (Fig. 57, $B/H = 46{\cdot}0$ m/$50{\cdot}1$ m) through the variable thickness of the Furens Dam in France (Fig. 58, $B/H = 49{\cdot}1$ m/$52{\cdot}0$ m) to the modern triangle with a vertical upstream face. The highest massive gravity dam is the Grande Dixence Dam in the Hérémence Valley in Switzerland with $B/H = 222$ m/285 m and a concrete volume of $5{\cdot}96 \times 10^6$ m^3.

The ratio of base width to height follows from the formula[33]

$$\frac{B}{H} = \left(\frac{1}{\gamma_c/\gamma_w - m}\right)^{1/2} = 0{\cdot}63 \text{ to } 0{\cdot}82 \text{ (average } 0{\cdot}71) \qquad (57)$$

where m denotes the ratio of the linearly distributed water pressure at the dam's base (uplift) to the hydrostatic pressure.

In 1925 *Vogt*[50] deduced the following three simple formulae for

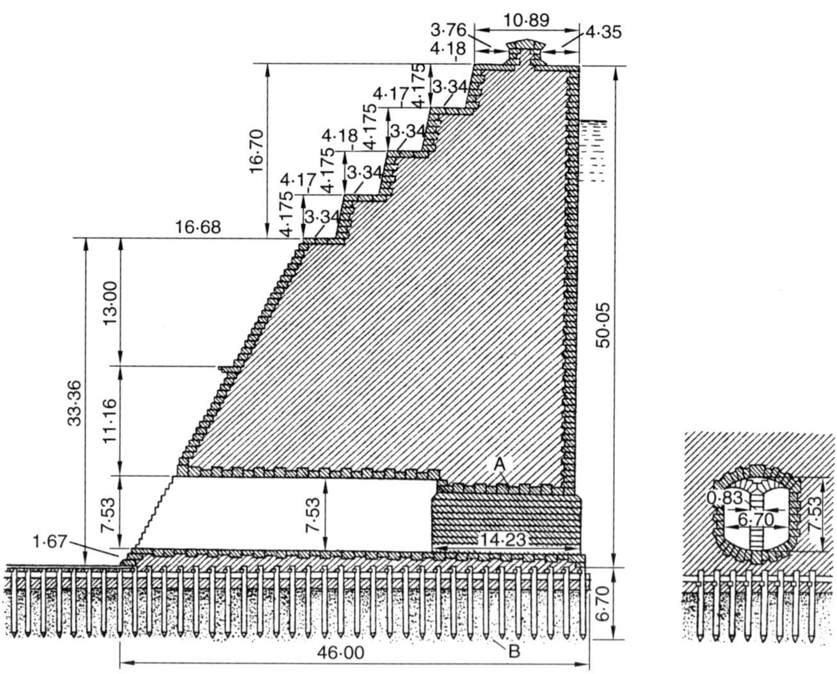

Fig. 57. Cross-section of Puentes gravity dam near Lorca, Spain, built from 1785 to 1791 and collapsed by leakage in 1802[91]

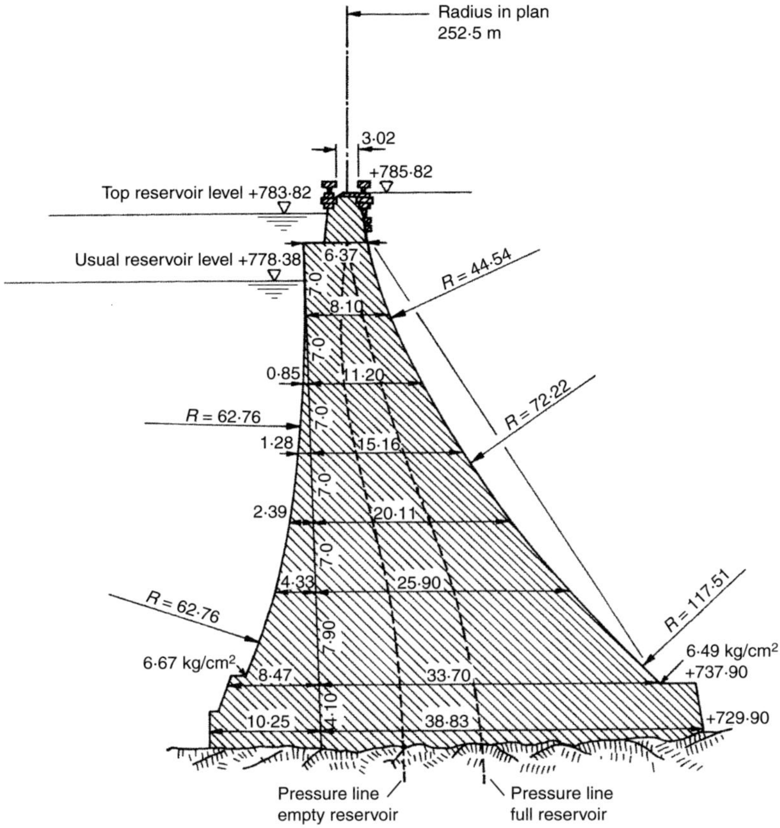

Fig. 58. Cross-section of Furens gravity dam near St. Etienne, France, built from 1861 to 1866[91]

ground deformations from *Boussinesq*'s theory of elastic semi-space of 1885 (Fig. 59).

(a) Normal displacement (settlement)

$$u = \frac{N}{E}\left(\frac{L}{B}\right)^{1/3} \tag{58}$$

(b) Tangential displacement

$$w = \frac{Q}{E}\left(\frac{L}{B}\right)^{1/3} \tag{59}$$

(c) Rotation

$$\alpha = \frac{18\ M}{\pi E B^2} \tag{60}$$

4.2.1. Two-dimensional case

In wide valleys it generally suffices to consider only the highest cross-section of gravity dams. The analysis of the usual triangular wedge can even be performed with closed formulae.[132]

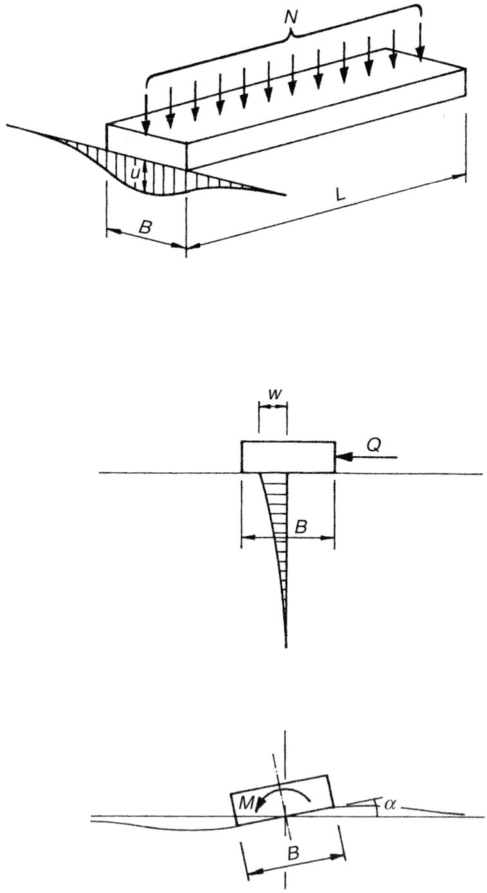

Fig. 59. Underground deformations[86]

4.2.1.1. Hydrostatic pressure

The crest displacement due to hydrostatic pressure is influenced by the bending M and shear Q deformations of both concrete

$$w_c = w_c^M + w_c^Q = \int_0^H \frac{M^0 M'}{E_c I} \mathrm{d}s + \int_0^H \frac{Q^0 Q'}{G_c A} \mathrm{d}s = \frac{\gamma_w H^5}{E_c B^3} + \frac{\gamma_w H^3}{4 G_c B} \tag{61}$$

and bedrock

$$w_r = w_r^M + w_r^Q = \frac{18 M^0 H}{\pi E_r B^2} + \frac{Q^0}{E_r}\left(\frac{L}{B}\right)^{1/3} = \frac{3\gamma_w H^4}{\pi E_r B^2} + \frac{\gamma_w H}{2E_r}\left(\frac{L}{B}\right)^{1/3} \tag{62}$$

and amounts finally to

$$w = w_c + w_r \tag{63}$$

4.2.1.2. Reservoir load

In the design of the 285 m high Grande Dixence Dam the influence of the enormous reservoir load was analysed by a model test.[36] The reservoir load causes both a *settlement of the valley floor* (Fig. 60) depending on the reservoir's depth H and width B[133]

$$\max\ u_r = 0{\cdot}39 \frac{\gamma_w H B}{E_r} \tag{64}$$

and a *widening of the valley* depending on the slope of the valley

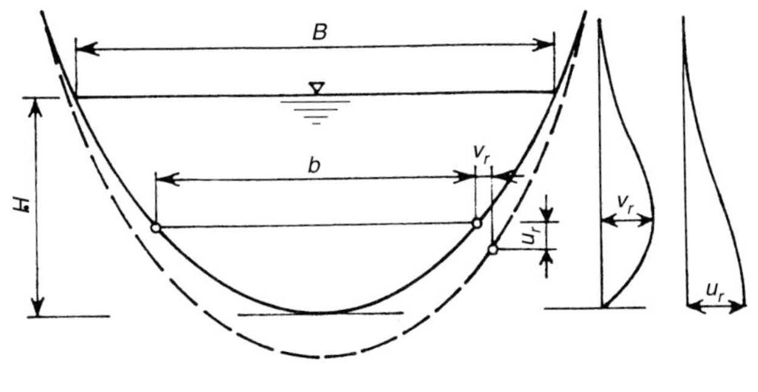

Fig. 60. Displacement of valley profile due to reservoir load[132]

flanks. In the lower quarter-point of the dam's height it amounts to

$$\max\ v_r = 0{\cdot}50\left(\frac{\gamma_w HB}{E_r}\right)\sin\beta\cos\beta \tag{65}$$

In the vertical plane of the dam's upstream face the rock deformations reach only 60% of the maximum values according to Equations (64) and (65).

The slope of the settlement depression below the dam equals the rotation of the valley floor and yields a negative (upstream) crest displacement (Fig. 61) of

$$w = \max\ u_r/2 \tag{66}$$

4.2.1.3. Changes in temperature

The difference in temperature between the upstream and downstream sides of the dam, even in a shallow depth of 1 m below the concrete surfaces, amounts to not more than a fraction of the surface value. For the difference in concrete temperature ΔT_c the crest displacement (Fig. 62) is given by

$$w^T = \frac{\omega \Delta T_c H}{2}\left(\frac{1}{\sin\beta\cos\beta} + \frac{1}{\tan\beta}\right) \tag{67}$$

4.2.1.4. Deformation moduli

Self-evidently, an analysis of dam deformations can be manipulated by the choice of appropriate deformation moduli for both concrete

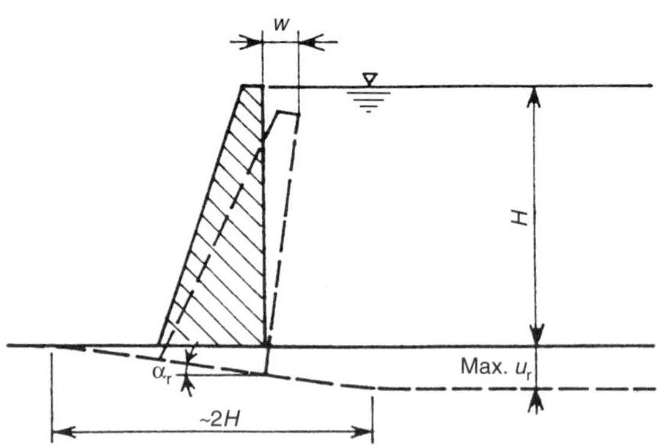

Fig. 61. Rotation of valley floor due to reservoir load[132]

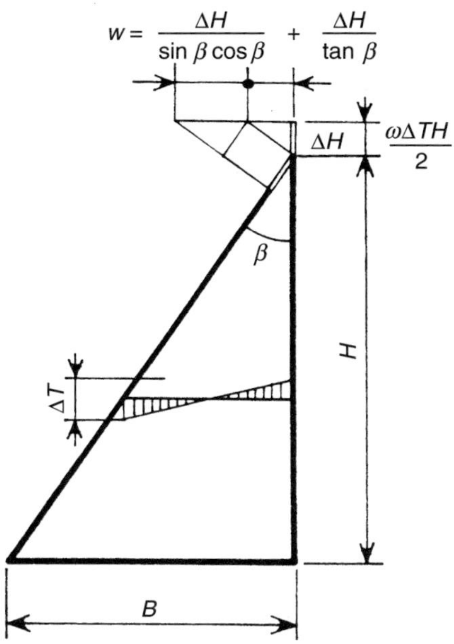

Fig. 62. Crest displacement of a gravity dam due to a linear temperature difference between upstream and downstream faces[132]

and bedrock. The usual assumption of $20\,000 < E_c = E_r < 40\,000$ MN/m^2 gives realistic values as experience shows.

4.2.1.5. Stresses

The usual assumption of a linear stress distribution across the dam section deviates from reality in the region of the dam's base.[33] This theoretical prediction (Fig. 10) is supported by photo-elastic model tests.[134] Whilst the downstream stress at the dam's base for a linear stress distribution and for the uplift ratio m amounts to[33]

$$\max\ \sigma_c^0 = \gamma_c H\left(\frac{1 + \gamma_c/\gamma_w - m}{\gamma_c/\gamma_w - m}\right) \tag{68}$$

the real stress depends on the deformation moduli for concrete and bedrock

$$\max\ \sigma_c = \max\ \sigma_c^0[1 + 0{\cdot}5(E_c/E_r)^{1/2}] \tag{69}$$

4.2.1.6. *Safety against sliding*

As several gravity dams actually failed by sliding on the bedrock (1911 Austin Dam in Pennsylvania, USA and 1928 St. Francis Dam in California, USA) the proof of a sufficient safety against sliding is of practical importance.

The deadload of the gravity dam of triangular cross-section

$$G = \gamma_c\left(\frac{BH}{2}\right) \tag{70}$$

the uplift at the dam's base

$$U = m\gamma_w\left(\frac{HB}{2}\right) \tag{71}$$

the hydrostatic pressure acting on the upstream face

$$W = \gamma_w\left(\frac{H^2}{2}\right) \tag{72}$$

and the adhesion of the concrete dam upon the bedrock

$$C = cB \tag{73}$$

give the condition of sliding

$$\frac{W - C}{G - U} < \tan\phi \tag{74}$$

if $\tan\phi$ denotes the coefficient of friction between the concrete dam and bedrock. The generally small value of $\tan\phi = 0{\cdot}7$ can, by artificial indentation, be increased to at least $\tan\phi = 1{\cdot}0$. The magnitude of adhesion, for which values of $0 < c < 3$ MN/m^2 are considered,[40] is subject to some controversy.

4.2.2. *Three-dimensional case*

As the measured crest displacements of straight gravity dams in narrow valleys are smaller than the predicted ones for the two-dimensional case, a portion of the hydrostatic load must be carried to the valley flanks.[135] This three-dimensional load-carrying action can be accounted for reliably not only by a finite element analysis but also by a simple grid analysis[136] if attention is paid to the fact that the existing vertical and often not grouted construction joints cannot transfer tensile stresses.

4.3. Numerical example: Schräh Dam (Fig. 63)

At the time of its construction (1922 to 1924) the Schräh Dam was the highest gravity dam in the world and it has been monitored since 1925.[101,134,137] Unfortunately, the thermometers installed refused further action after 30 years of service. It took 20 years for the hydration heat of the normal Portland cement (dosage 189 kg/m^3 of chuted concrete) to dissipate. Stored cubes, core-drilled cylinders and cut cubes showed in 1956 (at an age of 32 years) compressive strengths of 11·8 to 35·0 MN/m^2 (average 21·8 MN/m^2 (Ref. 101, p. 67)).

4.3.1. Two-dimensional analysis

Because of the statically ineffective narrow gorge in the lowest part of the Schräh Dam (Fig. 64) only an effective height of $H = 92{\cdot}5$ m is taken into account. Then, the crest displacement of the highest cantilever due to hydrostatic pressure is as follows.

Fig. 63. View of Schräh gravity dam in the Wägi Valley, Switzerland[42]

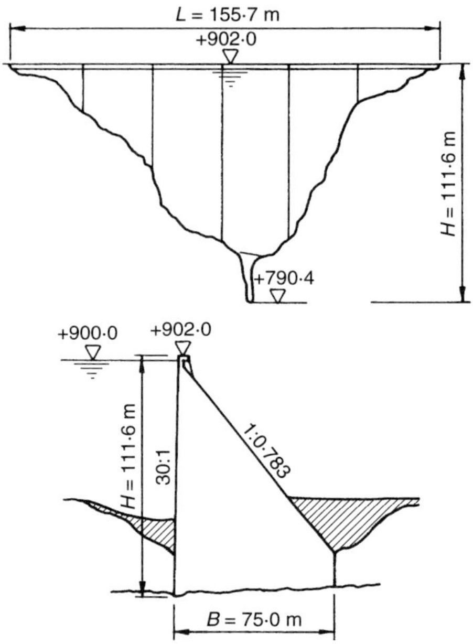

Fig. 64. Dimensions of Schräh Dam (the narrow portion of the gorge is 19 m deep)[42]

(a) Concrete alone, after Equation (61) with $E_c = 20\,000$ MN/m^2 and $G_c = 8000$ MN/m^2

$$w_c = \frac{0{\cdot}01 \times 92{\cdot}5^5}{20\,000 \times 75{\cdot}0^3} + \frac{0{\cdot}01 \times 92{\cdot}5^3}{4 \times 80000 \times 75{\cdot}0}$$
$$= 0{\cdot}0080 + 0{\cdot}0033 = 0{\cdot}0113 \text{ m}$$

(b) Bedrock alone, after Equation (62) with $E_r = E_c$

$$w_r = \frac{3 \times 0{\cdot}01 \times 92{\cdot}5^4}{\pi \times 20\,000 \times 75{\cdot}0^2} + \frac{0{\cdot}01 \times 92{\cdot}5^2}{2 \times 20\,000}$$
$$\times \left(\frac{155{\cdot}7}{75{\cdot}0}\right)^{1/3} = 0{\cdot}0062 + 0{\cdot}0027 = 0{\cdot}0089 \text{ m}$$

Then, from Equation (63)

$$w = 0{\cdot}0113 + 0{\cdot}0089 = 0{\cdot}0202 \text{ m}$$

Due to the reservoir load of $1{\cdot}47 \times 10^6$ MN, Equation (64) gives the settlement of the valley floor as

$$\max\ u_r = 0{\cdot}39 \times \frac{0{\cdot}01 \times 92{\cdot}5 \times 500}{20\,000} = 0{\cdot}0090 \text{ m}$$

Rotation of the valley floor (slope of the settlement depression) gives a negative (upstream) crest displacement of (Equation (66))

$$w = -0{\cdot}0090/2 = -0{\cdot}0045 \text{ m}$$

The total crest displacement due to hydrostatic pressure and reservoir load is then

$$\max\ w = 0{\cdot}0202 - 0{\cdot}0045 = 0{\cdot}0157 \text{ m}$$

For the drawdown level at El. 880 m (water depth 70·5 m) the corresponding crest displacement can be approximated as

$$\min\ w = 0{\cdot}0157\left(\frac{70{\cdot}5}{92{\cdot}5}\right)^4 = 0{\cdot}0053 \text{ m}$$

Then, the difference of crest displacements is

$$\Delta w^{\mathrm{W}} = 0{\cdot}0157 - 0{\cdot}0053 = 0{\cdot}0104 \text{ m}$$

Because of the dam's thickness (average value $t_m = 75{\cdot}0/2 = 37{\cdot}5$ m) the possible change of temperature from Equation (1) remains small

$$T_c = \frac{62}{37{\cdot}5 + 2{\cdot}4} = 1{\cdot}55\,°\text{C}$$

and the corresponding crest displacement follows from Equation (67) with $\omega = 10^{-5}$ and $\beta = 39°\ 13'$ as

$$w^{\mathrm{T}} = \mp\frac{10^{-5} \times 1{\cdot}55 \times 92{\cdot}5}{2}\left(\frac{1}{0{\cdot}632 \times 0{\cdot}775} + \frac{1}{0{\cdot}816}\right)$$
$$= \mp 0{\cdot}0023 \text{ m}$$

At the time of drawdown in spring the dam is still cold and for full reservoir in autumn it is still warm. Thus, the difference in crest displacements is reduced by the temperature changes to

$$\Delta w^{\mathrm{W+T}} = 10{\cdot}4 - 2 \times 2{\cdot}3 = 5{\cdot}8 \text{ mm}$$

This theoretical value is about twice the observed long-term value of 3·2 mm (Ref. 101, p. 70, Figures 4.2.2 – 6).

From Equations (68) and (69) the highest compressive stress at the dam's toe for $m = 0{\cdot}5$ and $E_r/E_c = 1$ is

$$\max\ \sigma_c = -0{\cdot}024 \times 92{\cdot}5\left(\frac{1 + 2{\cdot}4 - 0{\cdot}5}{2{\cdot}4 - 0{\cdot}5}\right)(1 + 0{\cdot}5)$$
$$= -5{\cdot}08 \text{ MN/m}^2$$

With the dam's deadload given by Equation (70)

$$G = 0{\cdot}024\left(\frac{75{\cdot}0 \times 92{\cdot}5}{2}\right) = 83{\cdot}25 \text{ MN/m}$$

the uplift from Equation (71)

$$U = 0{\cdot}5 \times 0{\cdot}01\left(\frac{92{\cdot}5 \times 75{\cdot}0}{2}\right) = 17{\cdot}34 \text{ MN/m}$$

the hydrostatic pressure from Equation (72)

$$W = 0{\cdot}01\left(\frac{92{\cdot}5^2}{2}\right) = 42{\cdot}78 \text{ MN/m}$$

and the adhesion from Equation (73)

$$C = 0{\cdot}1 \times 75{\cdot}0 = 7{\cdot}50 \text{ MN/m}$$

the safety against sliding given by Equation (74) for indentated bedrock with $\tan\phi = 1{\cdot}0$ is

$$s = 1{\cdot}0\left(\frac{83{\cdot}25 - 17{\cdot}34}{42{\cdot}78 - 7{\cdot}50}\right) = 1{\cdot}87$$

4.3.2. Three-dimensional dam action

Comparison of the long-term measured difference of crest displacements of $\Delta w = 3{\cdot}2$ mm for the last 40 years with the two-dimensional prediction of $\Delta w^{\mathrm{W+T}} = 5{\cdot}8$ mm shows that,

$[(5{\cdot}8-3{\cdot}2)/5{\cdot}8]\times 100=45\%$ of the hydrostatic load is deviated to the valley flanks. Then, the highest compressive stress at the dam's toe amounts only to about half the predicted value by two-dimensional analysis, max $\sigma_c=-0{\cdot}55\times 5{\cdot}08=-2{\cdot}79$ MN/m^2, whereas the lowest measured concrete strength is 11·8 MN/m^2. The safety against sliding is also much higher

$$s = 1{\cdot}0\left(\frac{83{\cdot}25 - 17{\cdot}34}{0{\cdot}55 \times 42{\cdot}78 - 7{\cdot}50}\right) = 4{\cdot}11 > 1{\cdot}87$$

than the two-dimensional prediction.

4.4. Numerical example: Grande Dixence Dam (Fig. 65)

This highest gravity dam to date was built in the Hérémence Valley in the Swiss Canton Valais from 1951 to 1962. Its design and analysis have previously been described in depth.[135] Monitoring of the dam's behaviour over several decades (Ref. 101, pp. 48 – 55 and Refs 134 and 137) permits a reliable comparison of measurement and prediction.

Fig. 65. View of the Grande Dixence gravity dam in the Hérémence Valley, Switzerland[42]

4.4.1. Two-dimensional analysis

The crest displacement of the dam's highest cross-section (Fig. 66) due to hydrostatic pressure is as follows.

(a) Concrete alone, after Equation (61) for $E_c = 25\ 000\ \mathrm{MN/m^2}$ and $G_c = 10\ 000\ \mathrm{MN/m^2}$

$$w_c = \frac{0{\cdot}01 \times 284^5}{25\,000 \times 222^3} + \frac{0{\cdot}01 \times 284^3}{4 \times 10\,000 \times 222}$$
$$= 0{\cdot}0675 + 0{\cdot}0258 = 0{\cdot}0933\ \mathrm{m}$$

(b) Bedrock alone, after Equation (62) for $E_r = E_c$

$$w_r = \frac{3 \times 0{\cdot}01 \times 284^4}{\pi \times 25\,000 \times 222^2} + \frac{0{\cdot}01 \times 284^2}{2 \times 25\,000}\left(\frac{910}{222}\right)^{1/3}$$
$$= 0{\cdot}0504 + 0{\cdot}0258 = 0{\cdot}0762\ \mathrm{m}$$

From Equation (63), $w = w_c + w_r$,

$$w = 0{\cdot}0933 + 0{\cdot}0762 = 0{\cdot}1695\,\mathrm{m}$$

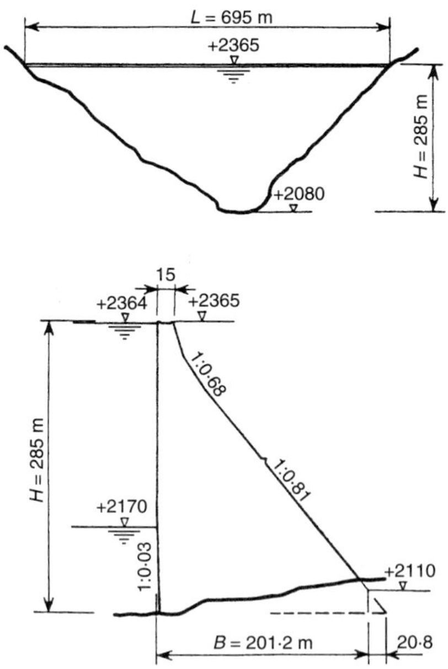

Fig. 66. Dimensions of Grande Dixence Dam[42]

The reservoir's capacity of 400×10^6 m^3 imposes the enormous load of $4{\cdot}0 \times 10^6$ MN. From Equation (64) the settlement of the valley floor is

$$\max\ u_r = 0{\cdot}39\left(\frac{0{\cdot}01 \times 284 \times 695}{25\,000}\right) = 0{\cdot}0308 \text{ m}$$

Rotation of the valley floor yields the negative (upstream) crest displacement of (Equation (66))

$$w_u = -0{\cdot}0308/2 = -0{\cdot}0154 \text{ m}$$

The total crest displacement due to hydrostatic pressure and reservoir load amounts finally to

$$w = 0{\cdot}1695 - 0{\cdot}0154 = 0{\cdot}1541 \text{ m}$$

Because of the dam's thickness (average $t_m = 222/2 = 111$ m) the change of concrete temperature in the Grand Dixence Dam according to Equation (1)

$$T_c = \frac{62}{111 + 2{\cdot}4} = \pm 0{\cdot}55\,^\circ\text{C}$$

remains very small. From Equation (67) the corresponding crest displacement with $\omega = 10^{-5}$ and $\beta = 37^\circ\ 55'$ is

$$w^T = \mp\frac{10^{-5} \times 0{\cdot}55 \times 285}{2}\left(\frac{1}{0{\cdot}615 \times 0{\cdot}789} + \frac{1}{0{\cdot}779}\right)$$
$$= \mp 0{\cdot}0026 \text{ m}$$

As the dam is still cold when the reservoir is drawn down in spring and still warm when the reservoir is full in autumn, the biggest possible difference in crest displacements cannot surpass the value of

$$\Delta w^{W+T} = 0{\cdot}1541 - 2 \times 0{\cdot}0026 = 0{\cdot}1489 \text{ m}$$

This theoretical prediction by a two-dimensional analysis is about twice the values measured in the following years.

1962 – 1965	0·066 m
1966 – 1970	0·081 m
1971 – 1975	0·085 m
1976 – 1980	0·078 m

Average $\Delta w = 0{\cdot}0775$ m.

The highest compressive stress at the dam's toe is given by Equations (68) and (69) for an uplift ratio of $m = 0{\cdot}3$ and ratio of deformation moduli of $E_r/E_c = 1{\cdot}0$ as

$$\max \sigma_c = -0{\cdot}025 \times 285\left(\frac{1 + 2{\cdot}5 - 0{\cdot}3}{2{\cdot}5 - 0{\cdot}3}\right)(1 + 0{\cdot}5)$$
$$= -15{\cdot}55 \text{ MN/m}^2$$

With the deadload of the highest cantilever from Fig. 66

$$G = 0{\cdot}025\left(\frac{193 \times 285}{2}\right) = 688 \text{ MN/m}$$

the uplift from Equation (71)

$$U = 0{\cdot}3 \times 0{\cdot}01\left(\frac{284 \times 193}{2}\right) = 82 \text{ MN/m}$$

the hydrostatic pressure from Equation (72)

$$W = 0{\cdot}01\left(\frac{284^2}{2}\right) = 403 \text{ MN/m}$$

and the adhesion from Equation (73)

$$C = 0{\cdot}1 \times 193 = 19 \text{ MN/m}$$

the safety against sliding on the upward sloping ($\tan 10^0 = 0{\cdot}176$) and indentated base of the dam ($\tan \phi = 1{\cdot}0$) is, according to Equation (74),

$$s = (1{\cdot}0 + 0{\cdot}176)\left(\frac{688 - 82}{403 - 19}\right) = 1{\cdot}86$$

4.4.2. Three-dimensional dam action

Comparison of the measured crest displacement of $\Delta w = 77{\cdot}5$ mm (average value for the years 1962 – 1980 (Ref. 101, p. 50)) with the two-dimensional prediction of $\Delta w^{W+T} = 148{\cdot}9$ mm shows that in reality $[(148{\cdot}9 - 77{\cdot}5)/148{\cdot}9] \times 100 = 48\%$ of the hydrostatic load is deviated to the valley flanks. Then, the highest compressive stress of the Grande Dixence Dam amounts only to half the value predicted by the two-dimensional analysis

$$\max\ \sigma_c = -(1 - 0{\cdot}48) \times 15{\cdot}55 = 8{\cdot}09\ \text{MN/m}^2$$

Also the safety against sliding is much higher

$$s = (1 + 0{\cdot}176)\left(\frac{688 - 82}{0{\cdot}52 \times 403 - 19}\right) = 3{\cdot}73 > 1{\cdot}86$$

than the value predicted by the two-dimensional analysis.

4.5. Numerical example: Koyna Dam (Fig. 67)

On 11 December 1967 the 103 m high Koyna gravity dam near Poona in India was subjected, by a strong-motion earthquake, to ground accelerations of $b_{Eh}/g = 0{\cdot}49$ in the transverse direction, $b_{Eh}/g = 0{\cdot}63$ in the longitudinal direction and $b_{Ev}/g = 0{\cdot}34$ in the vertical direction.[80,138]

4.5.1. Fictional crest displacement due to the deadload assumed to act horizontally in the transverse direction

Because of the special cross-section of this gravity dam the portion of crest displacement due to the deformation of concrete according

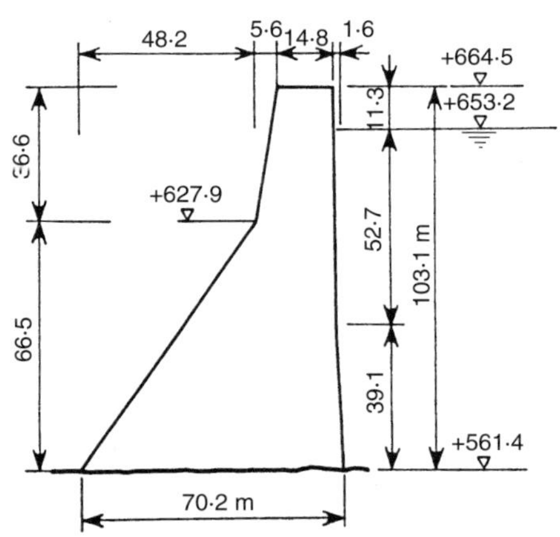

Fig. 67. Non-overflow cross-section of Koyna gravity dam in India before the earthquake of 11 December 1967[80]

to Equation (61) is evaluated numerically (Table 10). With

$$\gamma_c = 0{\cdot}026\ \mathrm{MN/m^3},\ \Delta H = 103{\cdot}1/6 = 17{\cdot}18\ \mathrm{m},$$
$$I_7 = 70{\cdot}2^3/12 = 28\,829\ \mathrm{m^4/m},\ E_{\mathrm{cdyn}} = 30\,000\ \mathrm{MN/m^2},$$
$$G_{\mathrm{cdyn}} = 12\,000\ \mathrm{MN/m^2}\ \text{and}\ Q' = 1\ \mathrm{m^{-1}}$$

it becomes

$$w_c = \frac{17{\cdot}18 \times 5\,008\,646}{3 \times 30\,000 \times 28\,829} + \frac{17{\cdot}18 \times 1029{\cdot}2}{3 \times 12\,000 \times 70{\cdot}2}$$
$$= 0{\cdot}0332 + 0{\cdot}0070 = 0{\cdot}0402\,\mathrm{m}$$

4.5.2. *Natural period of vibration*

Equation (36) gives the natural period of the Koyna Dam in the transverse direction as

$$T_1 = 1{\cdot}79\sqrt{0{\cdot}0402} = 0{\cdot}359\ \mathrm{s}$$

while the *in situ* measurement[80] was $T_1 = 0{\cdot}356$ s and a finite element analysis[138] produced a value of $T_1 = 0{\cdot}327$ s.

4.5.3. *Improved equivalent static load*

From the four multipliers given by Equations (37) – (40)

$$M_1 = 0{\cdot}5/0{\cdot}359 > 1{\cdot}0$$
$$M_2 = (2/4{\cdot}477)^2 = 0{\cdot}20$$
$$M_3 = 3/3 = 1{\cdot}0$$
$$M_4 = 1/(2 \times 0{\cdot}10) = 5 < 10$$

follows the ratio of the response acceleration of Koyna Dam to the ground acceleration in the transverse direction after Equation (42):

$$a_{\mathrm{Eh}}/b_{\mathrm{Eh}} = 1{\cdot}0 \times 0{\cdot}20 \times 1{\cdot}0 \times 5 = 1{\cdot}0$$

The seismic coefficient of the transverse oscillation is finally obtained from Equation (43):

$$C_s = 1{\cdot}0 \times 0{\cdot}49 = 0{\cdot}49$$

Table 10. Fictional crest displacement of Koyna Dam due to the deadload assumed to act horizontally in the transverse direction

i	t: m	g_0: MN/m²	G_0 : MN/m	Q^0 : MN/m	Q_0H : MN m/m	M^0 : MN m/m	M': m/m	I_7/I_i	k	$kM^0M'I_7/I_i$: MN m/m	A_7/A_i	$kQ^0Q'A_7/A_i$: MN/m²
1	14·8	0·385		0		0	0	106·7	1	0	4·74	0
			7·2		62							
2	17·4	0·452		7·2		62	17·18	65·7	4	279 924	4·03	116·1
			8·4		196							
3	20·1	0·523		15·6		258	34·37	42·6	2	755 508	3·49	108·9
			11·4		366							
4	31·0	0·806		27·0		624	51·55	11·6	4	1 492 558	2·26	244·1
			16·7		607							
5	43·9	1·141		43·7		1231	68·73	4·1	2	693 774	1·60	139·8
			22·5		944							
6	57·0	1·482		66·2		2175	85·92	1·9	4	1 420 258	1·23	325·7
			28·4		1381							
7	70·2	1·825				3556	103·10	1·0	1	366 624	1·00	94·6
Total										5 008 646		1029·2

4.5.4. Vertical stresses at El. 627·9 m

The equivalent static load of the dam portion above El. 627·9 m (Fig. 67) of

$$E_h = 0{\cdot}49\left(14{\cdot}8 + \frac{5{\cdot}6}{2}\right) \times 36{\cdot}6 \times 0{\cdot}026 = 8{\cdot}21\ \mathrm{MN/m}$$

with the lever arm of

$$a = \frac{36{\cdot}6}{2}\left(\frac{2 \times 14{\cdot}8 + 20{\cdot}4}{14{\cdot}8 + 20{\cdot}4}\right) = 17{\cdot}33\ \mathrm{m}$$

causes the bending moment of

$$M^{Eh} = \pm 8{\cdot}21 \times 17{\cdot}33 = \pm 142{\cdot}3\ \mathrm{MN\,m/m}$$

In addition to this acts the bending moment due to hydrostatic pressure of

$$M^{W} = -0{\cdot}01\left(\frac{25{\cdot}3^3}{6}\right)(1 \pm 0{\cdot}49) = \begin{matrix} -40{\cdot}22 \\ 13{\cdot}77 \end{matrix}\ \mathrm{MN\,m/m}$$

The deadload

$$G = 0{\cdot}026\left(14{\cdot}8 + \frac{5{\cdot}6}{2}\right)36{\cdot}6 = 16{\cdot}75\ \mathrm{MN/m}$$

with the eccentricity

$$e = \frac{20{\cdot}4}{2} - \frac{14{\cdot}8^2 + 14{\cdot}8 \times 20{\cdot}4 + 20{\cdot}4^2}{3(14{\cdot}8 + 20{\cdot}4)}$$
$$= 10{\cdot}20 - 9{\cdot}14 = 1{\cdot}06\ \mathrm{m}$$

causes normal forces ($a_{Ev}/g = 0{\cdot}34$) of

$$N = -16{\cdot}75(1 \pm 0{\cdot}34) = \begin{matrix} -22{\cdot}45 \\ -11{\cdot}06 \end{matrix}\ \mathrm{MN/m}$$

and corresponding bending moments of

$$M^{G} = \left|\begin{matrix} 22{\cdot}45 \\ 11{\cdot}06 \end{matrix}\right| \times 1{\cdot}06 = \begin{matrix} 23{\cdot}80 \\ 11{\cdot}72 \end{matrix}\ \mathrm{MN\,m/m}$$

The highest tensile stresses occur on the dam's upstream face as

$$\sigma_{xu} = -\frac{11{\cdot}06}{20{\cdot}4} + \frac{6(142{\cdot}3 + 40{\cdot}2 - 11{\cdot}7)}{20{\cdot}4^2}$$
$$= -0{\cdot}54 + 2{\cdot}46 = 1{\cdot}92 \text{ MN/m}^2$$

and on the downstream face as

$$\sigma_{xd} = -\frac{11{\cdot}06}{20{\cdot}4} + \frac{6(142{\cdot}3 + 13{\cdot}8 - 11{\cdot}7)}{20{\cdot}4^2}$$
$$= -0{\cdot}54 + 2{\cdot}08 = 1{\cdot}54 \text{ MN/m}^2$$

As these tensile stresses were larger than the tensile strength of concrete, cracks were inevitable.

4.6. Numerical example: Shasta Dam (Fig. 68)

From 1941 to 1945 this 183 m high gravity dam with a concrete volume of $4{\cdot}14 \times 10^6$ m^3 was built on the Sacramento River in the Central Valley of California.[139] Its abutment regions are slightly curved in plan. Its reservoir of $5{\cdot}2 \times 10^9$ m^3 is the backbone of California's water supply. Because of its size the Shasta Dam was generously equipped with instruments of proven value.

4.6.1. Stresses at El. 183 m

Concrete stresses were measured on 1 January 1943, 1944, 1945, 1946 and 1947 with nine-component strainmeters by Carlsson (Figs 69 to 71).[140] These instruments measured the electric resistance with an accuracy of 4×10^{-6}.

On 1 January 1946 the reservoir level was at El. 310 m and the depth of water to the measuring level at El. 183 m amounted to 127 m. The bending moment due to hydrostatic pressure

$$M^{W} = -0{\cdot}01\left(\frac{127^3}{6}\right) = -3414 \text{ MN m/m}$$

causes (wall thickness $t = 141{\cdot}5$ m) vertical stresses of

$$\sigma_x^{W} = \pm\frac{6{\cdot}3414}{141{\cdot}5^2} = \pm 1{\cdot}02 \text{ MN/m}^2$$

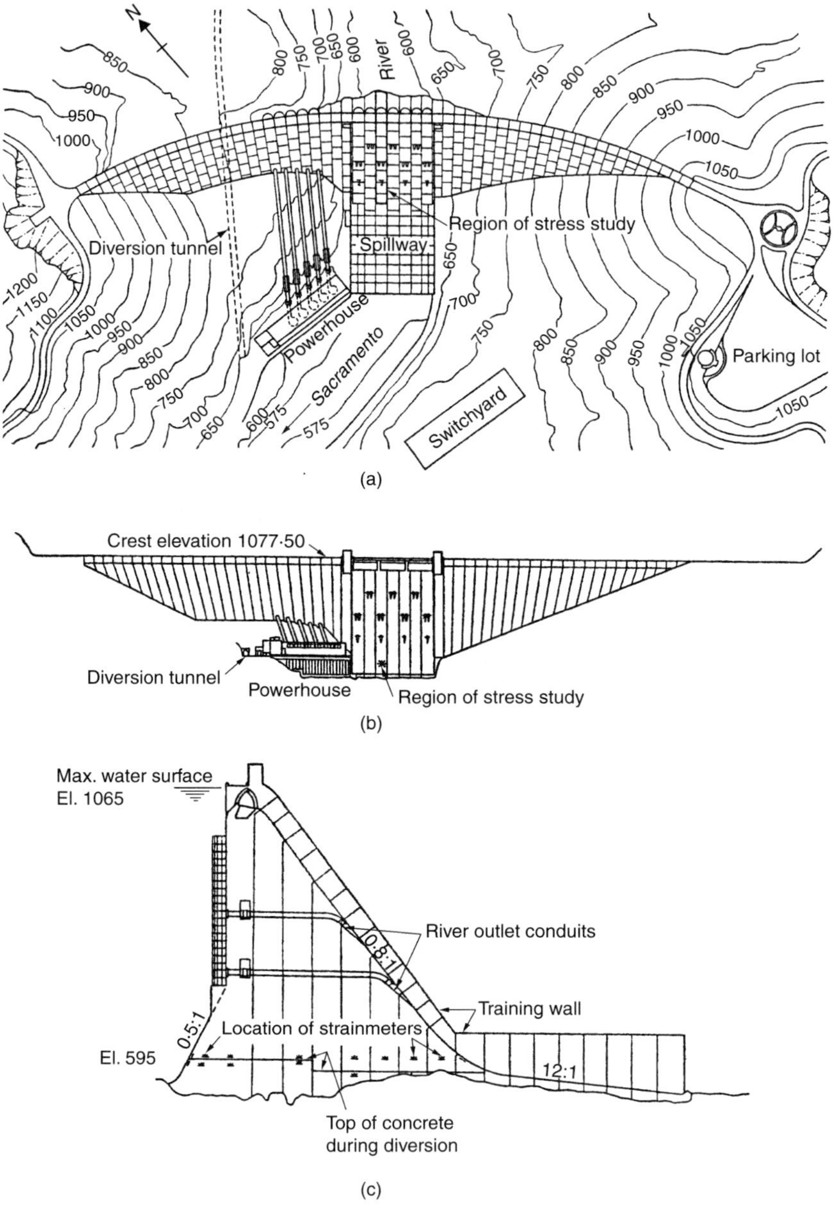

Fig. 68. Dimensions of Shasta gravity dam on the Sacramento River in the Central Valley of California:[140] (a) *plan,* (b) *downstream elevation,* (c) *spillway section row 13.*

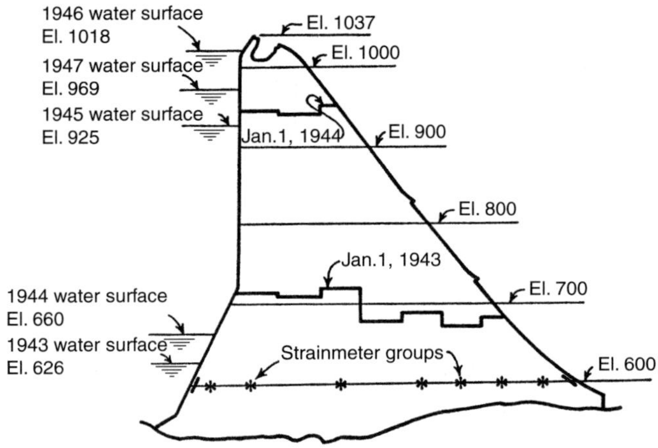

Fig. 69. Construction stages of Shasta Dam and impoundment[140]

The corresponding shear force of

$$Q^{\mathrm{W}} = 0{\cdot}01 \times \frac{127^2}{2} = 80{\cdot}65 \ \mathrm{MN/m}$$

causes an average shear stress of

$$\tau^{\mathrm{W}} = \frac{80{\cdot}65}{141{\cdot}5} = 0{\cdot}57 \ \mathrm{MN/m^2}$$

This predicted value is of the same magnitude as that observed (Fig. 70(*b*)).

For a linear stress distribution the vertical stresses due to deadload amount approximately to

$$\sigma_x^{\mathrm{G}} = -\frac{229}{141{\cdot}5}\left(1 \pm \frac{6 \times 16{\cdot}3}{141{\cdot}5}\right) = \begin{matrix} -2{\cdot}74 \\ -0{\cdot}50 \end{matrix} \ \mathrm{MN/m^2}$$

The resulting vertical stresses due to deadload and hydrostatic pressure

$$\sigma_x^{\mathrm{G+W}} = \begin{matrix} -2{\cdot}74 \\ -0{\cdot}50 \end{matrix} \pm 1{\cdot}02 = \begin{matrix} -1{\cdot}72 & (-1{\cdot}4) \\ -1{\cdot}52 & (-0{\cdot}7) \end{matrix} \ \mathrm{MN/m^2}$$

lie above those (in brackets) measured on 1 January 1946 (Fig. 70(*a*)).

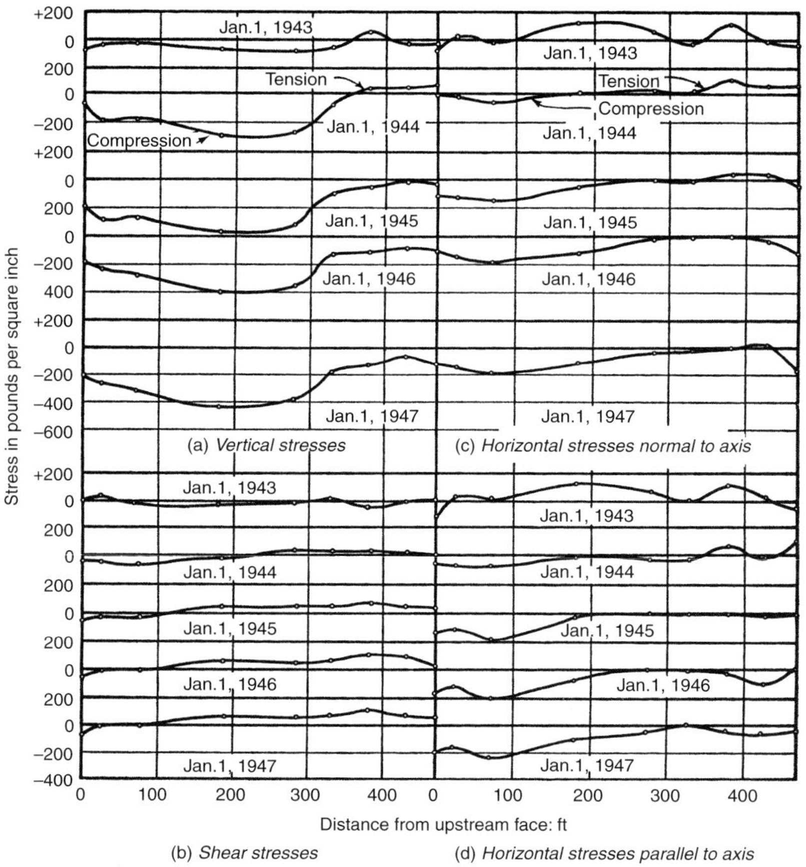

Fig. 70. Distribution of normal and shear stresses at El. 183 m of Shasta Dam[140]

As the Shasta Dam was constructed in blocks (columns with square cross-section of 15 m side length) their deadload stresses should be proportional to their heights, giving a maximum compressive stress of

$$\max\ \sigma_x^G = -0{\cdot}025 \times 133 = -3{\cdot}32\ \mathrm{MN/m^2}$$

In reality the smaller stress of $-2{\cdot}80\ \mathrm{MN/m^2}$ was measured (Fig. 70(*a*)) because friction between the individual blocks caused a certain redistribution of stresses from the dam's interior to its faces.

The highest principal stresses were expected in the region of the upstream third-point (Fig. 71(*d*)). With vertical stresses (linear stress

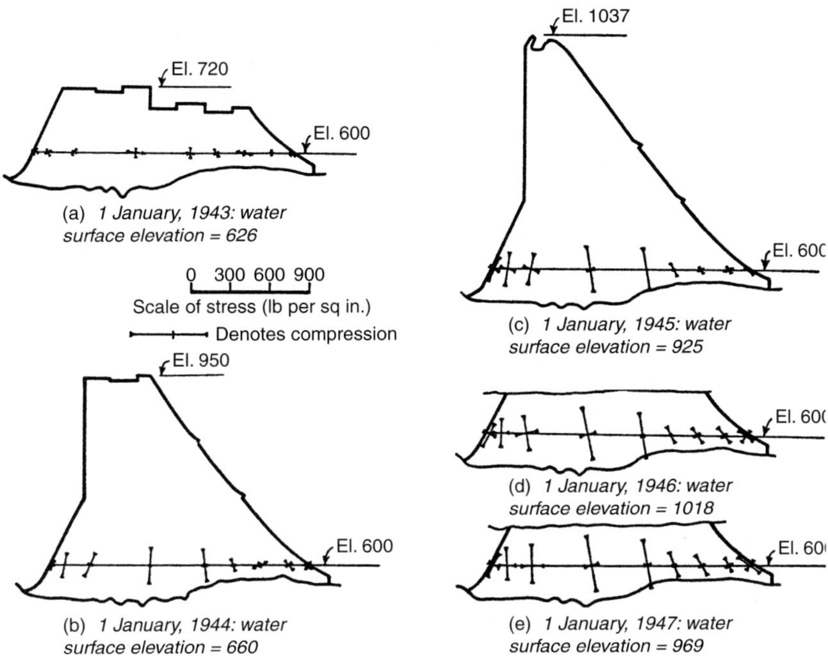

Fig. 71. Distribution of principal stresses at El. 183 m of Shasta Dam[140]

distribution presumed) due to deadload of $\sigma_x^G = -1{\cdot}99$ MN/m^2 and due to hydrostatic pressure of $\sigma_x^W = -0{\cdot}34$ MN/m^2 and average shear stress due to hydrostatic pressure of $\tau^W = 0{\cdot}57$ MN/m^2, the principal stresses amount to

$$\sigma_{1,2} = -\frac{2{\cdot}33}{2} \pm \left[\left(\frac{2{\cdot}33}{2}\right)^2 + 0{\cdot}57^2\right]^{1/2}$$

$$= -1{\cdot}17 \pm (1{\cdot}37 + 0{\cdot}32)^{1/2} = \begin{matrix} 0{\cdot}13 \\ -2{\cdot}47 \end{matrix} \ \text{MN/m}^2$$

whereas values of $\sigma_1 = 0{\cdot}83$ MN/m^2 (tension) and $\sigma_2 = -2{\cdot}85$ MN/m^2 (compression) were measured on 1 January 1946. With

$$\tan\ 2\alpha = \frac{2 \times 0{\cdot}57}{2{\cdot}33} = 0{\cdot}489 \quad \text{and} \quad \alpha = 13^\circ\ 02'$$

the deviation of the principal compressive stress from the vertical is of the same magnitude as the observed value of 12° 30′.

4.6.2. Temperature stresses

Figure 72 shows the measured air temperatures (below) and concrete stresses (curves 1 and 2 parallel and curve 3 normal to the surface) at a depth of 0·60 m below the concrete surface. The measured temperature stresses parallel to the surface amounted to $\sigma^T = 0{\cdot}7$ MN/m^2 (tension) and $\sigma^T = -6{\cdot}3$ MN/m^2 (compression) at concrete temperatures of $T = 11$ °C and $T = 30$ °C respectively. The stress difference of $\Delta\sigma^T = 7{\cdot}0$ MN/m^2 corresponds to a temperature difference of $\Delta T = 19$ °C. With the coefficient of thermal expansion for concrete of about $\omega = 10^{-5}$, Young's modulus of elasticity can be deduced from the above mentioned values:

$$E_c = \frac{7{\cdot}0}{19 \times 10^{-5}} = 37\,000 \text{ MN/m}^2$$

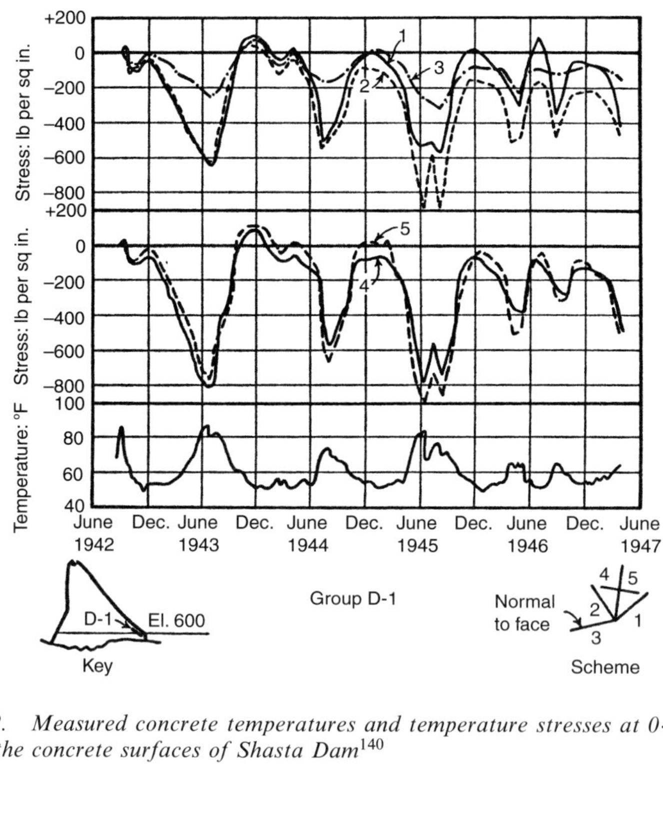

Fig. 72. Measured concrete temperatures and temperature stresses at 0·6 m depth below the concrete surfaces of Shasta Dam[140]

A real surprise was the measured stress difference normal to the concrete surface of $\Delta\sigma^{T} = 2{\cdot}1$ MN/m^2 (tension).

These measured temperature stresses make it clear why so many concrete dams possess damaged surfaces.

4.7. Numerical example: Mooser Dam (Fig. 73)

Constructed from 1951 to 1955 this 107 m high gravity dam with a slight curvature in plan ($R = 425$ m) has a crest length of 494 m and a concrete volume of 665 000 m^3.[141] The Mooser Dam rests on a narrow rock threshold between the 'Hoehenburg' and the 'Heidnische Kirche' (designations of localities) in the schist cover of the High Tauern Mountains in the Austrian Province of Salzburg. The Mooser Dam has served its purpose for many decades without any problems (Ref. 114, pp. 113 – 116). Because of its only slight curvature in plan, the Mooser Dam is analysed here as a straight gravity dam.

4.7.1. *Crest displacements*

The positive (downstream) crest displacement is as follows.

(a) Concrete alone, after Equation (61) with $E_c = 20\,000$ MN/m^2 and $G_c = 8000$ MN/m^2

$$w_c = \frac{0{\cdot}01 \times 107^5}{20\,000 \times 70^3} + \frac{0{\cdot}01 \times 107^3}{4 \times 8000 \times 70}$$
$$= 0{\cdot}0204 + 0{\cdot}0055 = 0{\cdot}0259 \text{ m}$$

(b) Bedrock alone, after Equation (62) for $E_r = E_c$

$$w_r = \frac{3 \times 0{\cdot}01 \times 107^4}{\pi \times 20\,000 \times 70^2} + \frac{0{\cdot}01 \times 107^2}{2 \times 20\,000}\left(\frac{494}{70}\right)^{1/3}$$
$$= 0{\cdot}0128 + 0{\cdot}0055 = 0{\cdot}0183 \text{ m}$$

Equation (63) then gives

$$w = 0{\cdot}0259 + 0{\cdot}0183 = 0{\cdot}0442 \text{ m}$$

Due to the reservoir load of 870 000 MN the negative (upstream) crest displacement is given by Equations (64) and (66) as

$$w_u = -\frac{0{\cdot}39}{2}\left(\frac{0{\cdot}01 \times 107 \times 1000}{20\,000}\right) = -0{\cdot}0104 \text{ m}$$

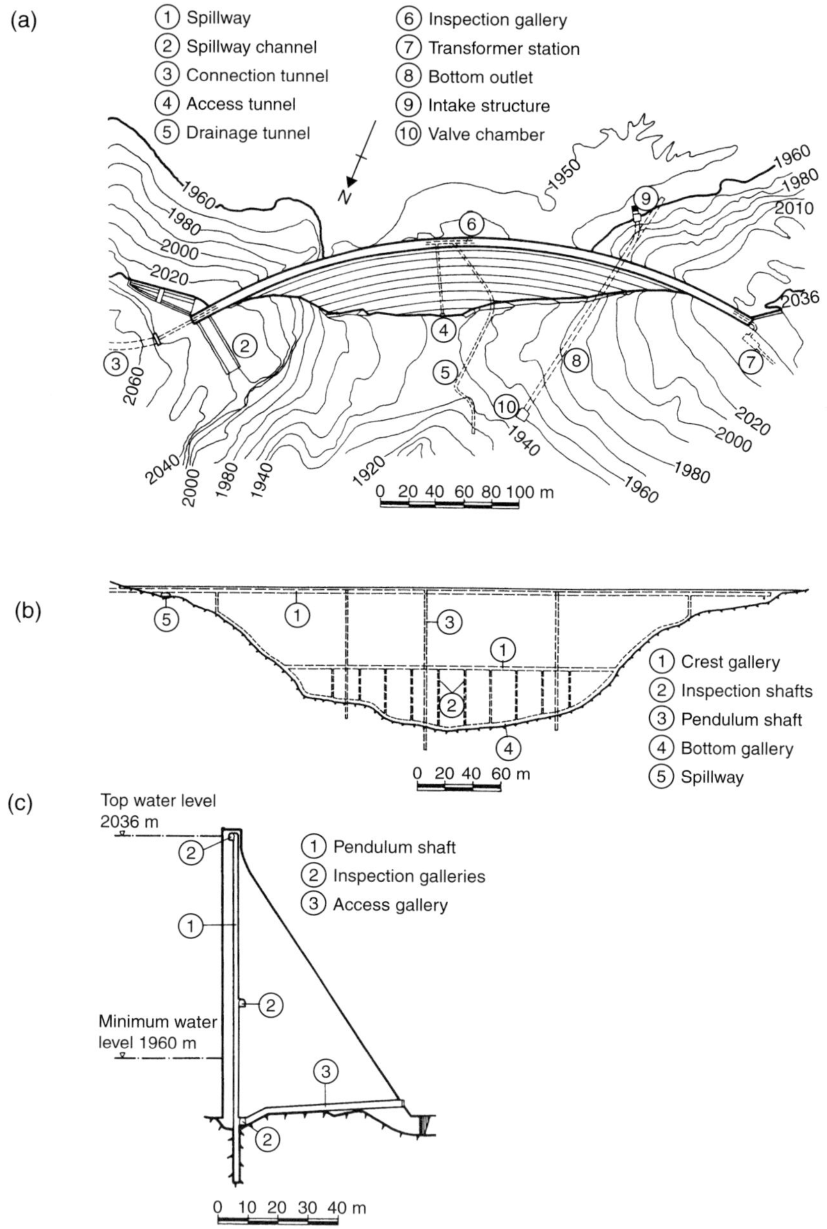

Fig. 73. Dimensions of Mooser gravity dam in the Kaprun Valley in Austria: (a) *plan,* (b) *downstream view and* (c) *crown cantilever*[114]

Thus, the total crest displacement due to hydrostatic pressure and reservoir load is, for a full reservoir,

$$\max\ w^{\mathrm{W}} = 0{\cdot}0442 - 0{\cdot}0104 = 0{\cdot}0338\ \mathrm{m}$$

If the reservoir level is lowered to El. 1960 m or by 76 m, the crest displacement can be approximated as

$$\min\ w^{\mathrm{W}} = 0{\cdot}0338\left(\frac{31}{107}\right)^4 = 0{\cdot}0002\ \mathrm{m}$$

The difference of crest displacements due to impounding then amounts to

$$\Delta w^{\mathrm{W}} = 0{\cdot}0338 - 0{\cdot}0002 = 0{\cdot}0336\ \mathrm{m}$$

The annual temperature variation of the Mooser Dam (warm in autumn with a full reservoir and cold in spring with an empty reservoir) is given by Equation (1) as

$$T_{\mathrm{c}} = \frac{62}{35 + 2{\cdot}4} = \pm 1{\cdot}7\,^{\circ}\mathrm{C}$$

With $\omega = 10^{-5}$ and $\beta = 33^{\circ}\ 12'$ the corresponding crest displacement is given by Equation (67)

$$\begin{aligned} w^{\mathrm{T}} &= \mp\frac{10^{-5}\times 1{\cdot}7\times 107}{2}\left(\frac{1}{0{\cdot}548\times 0{\cdot}837}+\frac{1}{0{\cdot}654}\right) \\ &= \mp 0{\cdot}0034\ \mathrm{m} \end{aligned}$$

and the difference of crest displacements due to temperature variations amounts to

$$\Delta w^{\mathrm{T}} = 2\times 0{\cdot}0034 = 0{\cdot}0068\ \mathrm{m}$$

As the crest displacements due to temperature changes oppose those due to the filling and emptying of the reservoir, the annual difference of crest displacements cannot exceed

$$\Delta w^{\mathrm{W+T}} = 0{\cdot}0336 - 0{\cdot}0068 = 0{\cdot}0268\ \mathrm{m}$$

The measured values decreased from $\Delta w = 32$ mm initially to 24 mm in the course of four decades. This coincidence of observation and elementary analysis is very satisfactory.

4.7.2. Stresses

For an uplift ratio of $m = 0{\cdot}5$ and for a ratio of deformation moduli $E_r/E_c = 1{\cdot}0$ the highest compressive concrete stress at the dam's toe is given by Equations (68) and (69) as

$$\max\ \sigma_c = -0{\cdot}025 \times 107\left(\frac{1 + 2{\cdot}5 - 0{\cdot}5}{2{\cdot}5 - 0{\cdot}5}\right)(1 + 0{\cdot}5)$$
$$= -6{\cdot}02\ \mathrm{MN/m^2}$$

4.7.3. Safety against sliding

With the deadload from Equation (70)

$$G = 0{\cdot}025\left(\frac{70 \times 107}{2}\right) = 93{\cdot}6\ \mathrm{MN/m}$$

the uplift from Equation (71)

$$U = 0{\cdot}5 \times 0{\cdot}01\left(\frac{107 \times 70}{2}\right) = 18{\cdot}7\ \mathrm{MN/m}$$

the hydrostatic pressure from Equation (72)

$$W = 0{\cdot}01\left(\frac{107^2}{2}\right) = 57{\cdot}3\ \mathrm{MN/m}$$

and the adhesion after Equation (73)

$$C = 0{\cdot}1 \times 70 = 7{\cdot}0\ \mathrm{MN/m}$$

the safety against sliding is achieved by a good indentation of the dam's base in the bedrock, and this is given by Equation (74) as

$$s = 1{\cdot}0\left(\frac{93{\cdot}6 - 18{\cdot}7}{57{\cdot}3 - 7{\cdot}0}\right) = 1{\cdot}49$$

In reality, this value is enhanced by the dam's curvature in plan.

4.8. Gravity dams with cavities

A gaping longitudinal crack (Fig. 74) can often be observed in massive gravity dams independent of their size. To counteract this danger four gravity dams with base caverns have already been built in Austria (1958 Big and Small Muehldorfer Lakes,[142] Hochalm Lake and 1973 New Tauernmoos Dam). An additional advantage of such dams is the restriction of uplift to the dam's heel because the base cavern acts as a drainage point.

4.8.1. Displacements and stresses

The displacements of gravity dams with base caverns are only slightly larger than those of massive gravity dams. In the case of hydrostatic pressure it suffices to enhance the second term of Equation (61) by a half, if the base cavern covers the middle third of the dam's base (Fig. 75) as is customary.

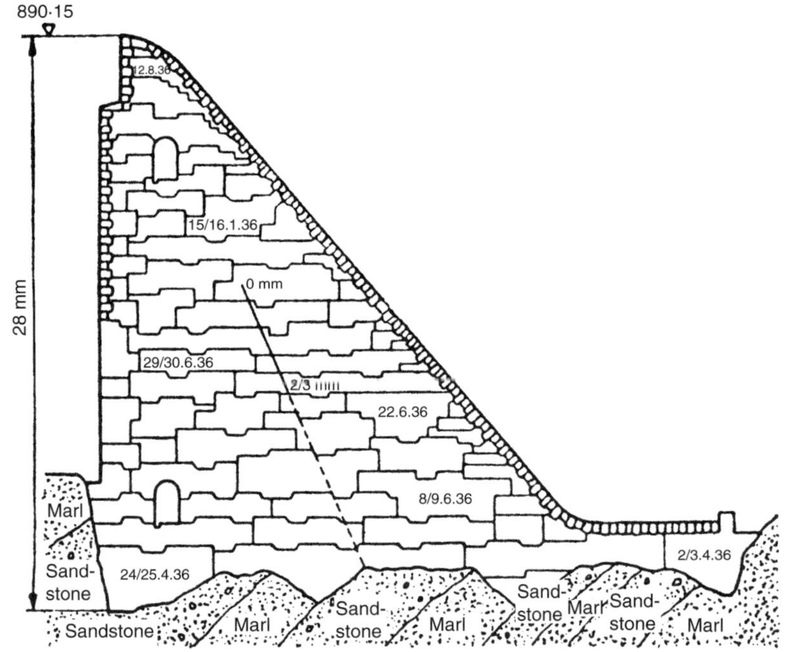

Fig. 74. 'In den Schlagen' gravity dam on Lake Sihl in Switzerland with a longitudinal crack[134]

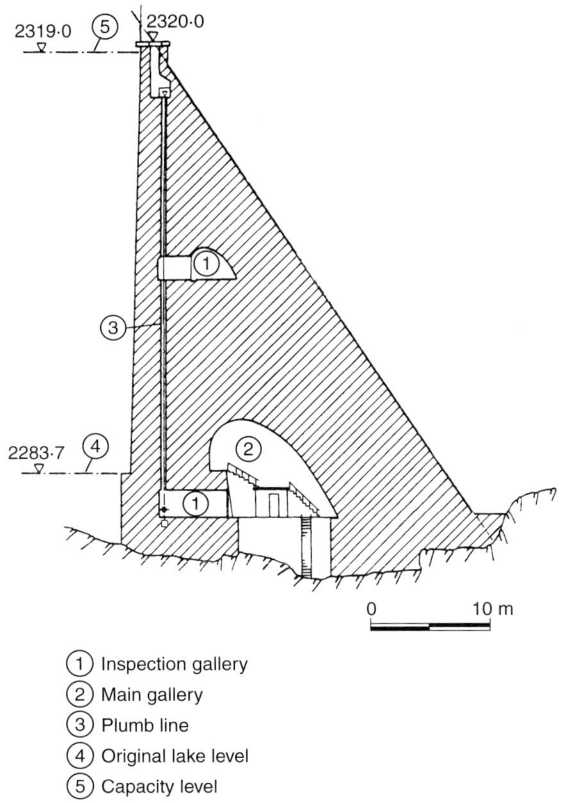

Fig. 75. Cross-section of gravity dam with base cavern damming the Big Mühldorfer Lake in Austria[114]

The area of the dam's base approximates in this case to

$$A = \tfrac{2}{3}B < A_0 = B \tag{75}$$

and the section modulus to

$$S = \frac{26}{27} \times \frac{B^2}{6} < S_0 = \frac{B^2}{6} \tag{76}$$

From a comparison with values of A_0 and S_0 for the base of a massive gravity dam, it follows that the shear stresses rise by 50% but the bending stresses increase by only $1/27 = 3{\cdot}7\%$.

4.8.2. Safety against sliding

The dam's deadload is reduced to about

$$G = \gamma_c B\left(\frac{H}{2} - \frac{B}{9}\right) \tag{77}$$

and the uplift to

$$U = m\gamma_w \frac{HB}{6} \tag{78}$$

The hydrostatic pressure on the upstream face remains the same as for a massive gravity dam and the adhesion is reduced to

$$C = \tfrac{2}{3}cB \tag{79}$$

The condition of sliding is identical to that for massive gravity dams.

4.9 Numerical example: Big Mühldorfer Lake Dam (Fig. 75)

Constructed from 1954 to 1957 this first gravity dam with a base cavern has a height of 46·5 m and a total base width (including the cavern) of 31·3 m. The crest length amounts to 433 m. The dam is slightly curved in plan ($R = 2630$ m) and rests on a rock threshold of eye gneiss. The reduction in concrete volume as compared to a massive gravity dam amounts to 10%.

4.9.1 Crest displacements

The positive (downstream) crest displacement due to hydrostatic pressure is as follows.

(a) Concrete alone, after Equation (61) with $E_c = 20\,000$ MN/m^2 and $G_c = 8000$ MN/m^2

$$w_c = \frac{0{\cdot}01 \times 45{\cdot}5^5}{20\,000 \times 31{\cdot}3^3} + \frac{3 \times 0{\cdot}01 \times 45{\cdot}5^3}{8 \times 8000 \times 31{\cdot}3}$$
$$= 0{\cdot}0032 + 0{\cdot}0014 = 0{\cdot}0046 \text{ m}$$

(b) Bedrock alone, after Equation (62) with $E_r = E_c$ to

$$w_r = \frac{3 \times 0{\cdot}01 \times 45{\cdot}5^4}{\pi \times 20\,000 \times 31{\cdot}3^3} + \frac{0{\cdot}01 \times 45{\cdot}5^2}{2 \times 20\,000}\left(\frac{445}{31{\cdot}3}\right)^{1/3}$$
$$= 0{\cdot}0021 + 0{\cdot}0012 = 0{\cdot}0033\,\text{m}$$

Equation (63) then yields:

$$w = 0{\cdot}0046 + 0{\cdot}0033 = 0{\cdot}0079\text{ m}$$

The reservoir contains $7{\cdot}9 \times 10^6$ m^3 and causes a negative (upstream) crest displacement of (Equations (64) and (66))

$$w_u = -\frac{0{\cdot}39}{2}\left(\frac{0{\cdot}01 \times 45{\cdot}5 \times 433}{20\,000}\right) = -0{\cdot}0019\text{ m}$$

The total crest displacement due to hydrostatic pressure and reservoir load amounts finally to

$$w^W = 0{\cdot}0079 - 0{\cdot}0019 = 0{\cdot}0060\text{ m}$$

This predicted value compares favourably with measurements taken over 30 years of service (Ref. 114, p. 127, Fig. 3) of about $\Delta w =$ 5·5 – 7·5 mm.

For an average wall thickness of $t_m = 16{\cdot}0$ m the dam's variation's of temperature are given by Equation (1) as

$$T_c = \frac{62}{16{\cdot}0 + 2{\cdot}4} = \pm 3{\cdot}4\ ^\circ\text{C}$$

and the corresponding crest displacements with $\omega = 10^{-5}$ and $\beta = 35^\circ\ 30'$ is given by Equation (67) as

$$w^T = \mp\frac{10^{-5} \times 3{\cdot}4 \times 46{\cdot}5}{2}\left(\frac{1}{0{\cdot}581 \times 0{\cdot}814} + \frac{1}{0{\cdot}713}\right)$$
$$= \mp 0{\cdot}0028\text{ m}$$

4.9.2. Stresses

Equations (68) and (69) are used to obtain the highest compressive stress at the dam's toe. With $m = 0{\cdot}8$ (measured) and $E_r/E_c = 1{\cdot}0$ this is

$$\max\ \sigma_c = -0{\cdot}025 \times 46{\cdot}5\left(\frac{1+2{\cdot}5-0{\cdot}8}{2{\cdot}5-0{\cdot}8}\right)(1+0{\cdot}5) \times \frac{27}{26}$$
$$= -2{\cdot}9\ \text{MN/m}^2$$

The designer's original analysis produced a value of $-5{\cdot}0\ \text{MN/m}^2$.

4.9.3. Safety against sliding

With the four auxiliary values obtained from Equations (72) and (77) – (79):

$$W = 0{\cdot}01\left(\frac{45{\cdot}5^2}{2}\right) = 10{\cdot}35\ \text{MN/m}$$
$$G = 0{\cdot}025 \times 31{\cdot}3\left(\frac{46{\cdot}5}{2} - \frac{31{\cdot}3}{9}\right) = 15{\cdot}47\ \text{MN/m}$$
$$U = 0{\cdot}8 \times 0{\cdot}01\left(\frac{45{\cdot}5 \times 31{\cdot}3}{6}\right) = 1{\cdot}90\ \text{MN/m}$$
$$C = \tfrac{2}{3} \times 0{\cdot}1 \times 31{\cdot}3 = 2{\cdot}09\ \text{MN/m}$$

the safety against sliding upon the well-indented bedrock ($\tan\phi = 1{\cdot}0$) is given by Equation (74) as

$$s = 1{\cdot}0\left(\frac{15{\cdot}47 - 1{\cdot}90}{10{\cdot}35 - 2{\cdot}09}\right) = 1{\cdot}64$$

4.10 Buttress dams

Despite the abundance of shapes of buttress dams (see Section 4.1) their analyses are similar.

4.10.1. Deadload

With the cross-sectional areas of A_1 at the crest, A_2 at midheight and A_3 at foundation level, the concrete volume is given by Simpson's rule as

$$V = \frac{H}{6}(A_1 + 4A_2 + A_3) \tag{80}$$

and the deadload of one buttress compartment is

$$G = \gamma_c V \tag{81}$$

4.10.2. Hydrostatic pressures

The components of hydrostatic pressure on the dam's inclined upstream face (Fig. 76) are given in the horizontal direction

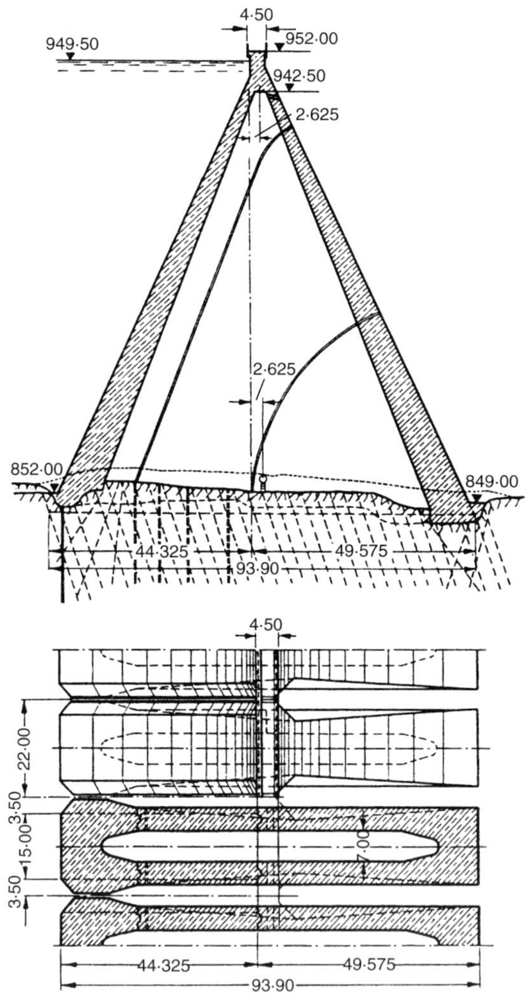

(a)

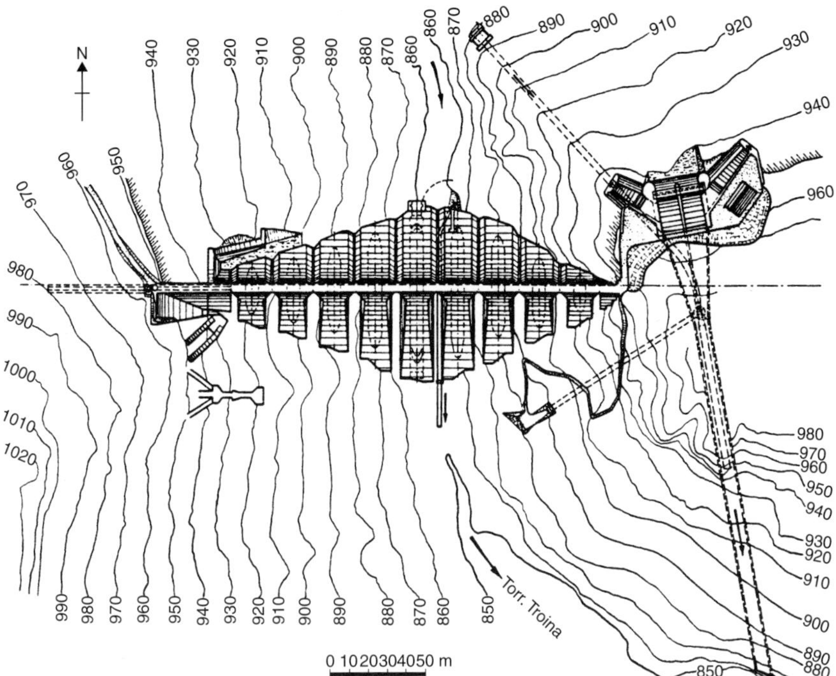

(b)

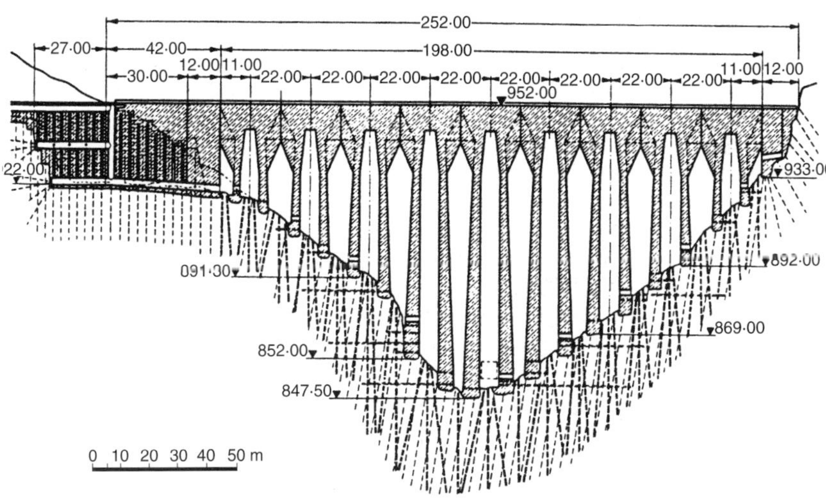

(c)

Fig. 76 (facing and above). Hollow buttress dam Ancipa in Sicily: (a) *cross-section,* (b) *plan and* (c) *longitudinal section*[131]

as

$$W_h = \gamma_w b\left(\frac{(H-f)^2}{2}\right) \tag{82}$$

and in the vertical direction for the dam's upstream slope of 1 in n as

$$W_v = nW_h \tag{83}$$

4.10.3. Uplift

Uplift can occur only under the buttress head (thickened upstream portion) as the empty space between the individual buttresses acts as drainage. Assuming a linear decrease of uplift from its maximum value of $m\gamma_w H$ at the water face to zero at the downstream end of the buttress head, the total uplift is given by

$$U = m\gamma_w\left(\frac{Hd}{2}\right) \tag{84}$$

where d is the length of the buttress head in its horizontal projection.

4.10.4. Adhesion

The dam's adhesion to the bedrock is approximated by

$$C = cA_3 \tag{85}$$

4.10.5. Stresses at foundation level

For a known moment of inertia of the base section and distances a_u from the upstream face and a_d from the downstream face to the centre-line, the section moduli are given for the upstream face as $S_u = I/a_u$ and for the downstream face as $S_d = I/a_d$. The normal force at foundation level

$$N = G + W_v - U \tag{86}$$

and the bending moment due to both horizontal and vertical loads with their eccentricities e

$$M = W_h\left(\frac{H-f}{3}\right) - Ge_G - W_v e_W + Ue_U \tag{87}$$

yield vertical stresses on the upstream face as

$$\sigma_u = \frac{N}{A_3} + \frac{M}{S_u} \tag{88a}$$

and on the downstream face as

$$\sigma_d = \frac{N}{A_3} - \frac{M}{S_d} \tag{88b}$$

For a slope of $\tan \beta = n$ the normal stresses parallel to the inclined upstream and downstream faces of the dam amount finally to

$$\max\ \sigma_u = \sigma_u / \cos \beta_u \tag{89a}$$

$$\max\ \sigma_d = \sigma_d / \cos \beta_d \tag{89b}$$

The shear stress of the buttress is given by

$$\tau = W_h / b_0 B \tag{90}$$

4.10.6. Shrinkage and temperature stresses

As thinner portions of the buttress section cool more quickly and shrink more than the thicker ones (buttress heads) tensile stresses are caused in the buttress, especially near the bedrock which hinders the shortening of concrete[143]

$$\sigma_h = [\varepsilon_s + \omega(T_1 + T_2)]E_c \tag{91}$$

where ε_s denotes the specific shrinkage of concrete, ω is its coefficient of thermal expansion, T_1 is the decrease of concrete temperature due to the dissipating hydration heat of the cement, and T_2 is the variation of concrete temperature due to a variation of air temperature after Equation (1).

4.10.7. Safety against sliding

The condition of sliding

$$\frac{W_h - C}{G + W_v - U} > \tan\ \phi \tag{92}$$

differs from Equation (74) only by the vertical component of hydrostatic pressure acting on the buttress dam's inclined upstream face.

4.10.8. Crest displacements

The crest displacements of a buttress dam due to hydrostatic pressure consist of the portions for concrete after Equation (61) and for bedrock after Equation (62). From their sum the negative (upstream) crest displacement due to the reservoir load given by Equation (66) must be subtracted.

4.11. Numerical example: Ancipa Dam (Fig. 76)

The hollow buttress dam Ancipa in Sicily,[131] built from 1951 to 1954, is, at a height of $H = 111{\cdot}5$ m, the highest such dam built to date (Fig. 76). The relevant section values of the highest buttress compartment are given in Table 11.

4.11.1. Deadload

The deadload of the highest buttress compartment is given by Equations (80) and (81) as

$$G = 0{\cdot}025\left(\frac{107{\cdot}5}{6}\right)(99 + 4{\cdot}557 + 1276) = 1614 \text{ MN}$$

4.11.2. Hydrostatic pressure

The two components of hydrostatic pressure are given by Equations (82) and (83) as

$$W_{\mathrm{h}} = 0{\cdot}01\left(\frac{105{\cdot}0^2}{2}\right)22{\cdot}0 = 1213 \text{ MN}$$

$$W_{\mathrm{v}} = 0{\cdot}45 \times 1213 = 546 \text{ MN}$$

Table 11. Buttress compartment parameters of Ancipa Dam

Cross-section	A: m^2	I: m^4
Crest	99	167
Midheight	557	141 983
Base	1276	1 151 737

4.11.3. Uplift

With an uplift ratio of $m = 1{\cdot}0$ (possible maximum) the uplift is given by Equation (84):

$$U = 1{\cdot}0 \times 0{\cdot}01 \times 22{\cdot}0\left(\frac{105{\cdot}0 \times 10{\cdot}0}{2}\right) = 116 \text{ MN}$$

4.11.4. Adhesion

With $c = 0{\cdot}1$ MN/m^2 Equation (85) yields an adhesion of

$$C = 0{\cdot}1 \times 1276 = 128 \text{ MN}$$

4.11.5. Safety against sliding

From Equation (92) the safety against sliding for the well-indented base of the Ancipa Dam is

$$s = 1{\cdot}0\left(\frac{1164 + 546 - 116}{1213 - 128}\right) = 1{\cdot}88$$

4.11.6. Stresses at foundation level

For the normal force of (Equation (86))

$$N = 1614 + 546 - 116 = 2044 \text{ MN}$$

and the bending moment of (Equation (87))

$$M = 1213\left(\frac{105{\cdot}0}{3}\right) - 1614 \times 2{\cdot}62 - 546 \times 29{\cdot}55 + 116$$
$$\times 39{\cdot}33 = 42\,455 - 4229 - 16\,134 + 4562 = 26\,654 \text{ MNm}$$

the vertical stresses are given by Equations (88*a*) and (88*b*) as

$$\sigma_{\mathrm{u}} = -\frac{2044}{1276} + \frac{26\,654}{25\,984} = -1{\cdot}60 + 1{\cdot}03 = -0{\cdot}57 \text{ MN/m}^2$$

$$\sigma_{\mathrm{d}} = -\frac{2044}{1276} - \frac{26\,654}{23\,232} = -1{\cdot}60 - 1{\cdot}15 = -2{\cdot}75 \text{ MN/m}^2$$

and the maximum stresses of the inclined dam faces by Equations (89*a*) and (89*b*) as

$$\max\ \sigma_u = -\frac{0 \cdot 57}{0 \cdot 912} = -0 \cdot 62\ \text{MN/m}^2$$

$$\max\ \sigma_d = -\frac{2 \cdot 75}{0 \cdot 912} = -3 \cdot 02\ \text{MN/m}^2$$

The shear stress is given by Equation (90):

$$\tau = \frac{1213}{2 \times 5 \cdot 50 \times 93 \cdot 9} = 1 \cdot 17\ \text{MN/m}^2$$

4.11.7. Shrinkage and temperature stresses

From Equation (91) with the specific shrinkage of concrete of $\varepsilon_s = -10 \times 10^{-5}$, the temperature decrease due to the dissipating hydration heat of cement of $T_1 = -12\ °C$, the decrease of concrete temperature due to the decrease of air temperature after Equation (1) of $T_2 = 62/(18 \cdot 0 + 2 \cdot 4) = -3 \cdot 0\ °C$, and Young's modulus of elasticity for concrete of $E_c = 20\ 000\ \text{MN/m}^2$, the horizontal tensile stress in the buttress is

$$\sigma_h = (10 + 12 + 3) \times 10^{-5} \times 20\,000 = 5 \cdot 00\ \text{MN/m}^2$$

4.11.8. Principal stresses

With a vertical stress of $\sigma_x = -1 \cdot 60\ \text{MN/m}^2$, a horizontal stress of $\sigma_y = 5 \cdot 00\ \text{MN/m}^2$ and a shear stress $\tau_{xy} = 1 \cdot 17\ \text{MN/m}^2$, the principal stresses at the base section's centre-line are computed as

$$\sigma_{1,2} = \frac{\sigma_x + \sigma_y}{2} \pm \left[\left(\frac{\sigma_x - \sigma_y}{2}\right)^2 + \tau_{xy}^2\right]^{1/2}$$

$$= \frac{-1 \cdot 60 + 5 \cdot 00}{2} \pm \left[\left(\frac{-1 \cdot 60 - 5 \cdot 00}{2}\right)^2 + 1 \cdot 17^2\right]^{1/2}$$

$$= 1 \cdot 70 \pm (10 \cdot 89 + 1 \cdot 37)^{1/2} = \begin{matrix} 5 \cdot 20 \\ -1 \cdot 80 \end{matrix}\ \text{MN/m}^2 \qquad (93)$$

Their deviation from the vertical is

$$\tan\ 2\alpha = \frac{2\tau_{xy}}{\sigma_x - \sigma_y} = \frac{2 \times 1{\cdot}17}{-1{\cdot}60 - 5{\cdot}00} = -0{\cdot}3545 \tag{94}$$

and $\alpha = 9^\circ\,46'$ respectively.

Expected cracking of the Ancipa Dam was counteracted with two joints of dilatation (Fig. 76(*a*)) whose traces are trajectories of the principal compressive stresses.

4.12. Multiple arch dams

Although the first multiple arch dam was built in India about two centuries ago (Meer Allum Dam in 1800) and such dams were often built in the beginning of the twentieth century, this type of dam is no longer in favour since the rupture of the Gleno Dam in Italy on 1 December 1923.[93,144] A rare exception was the construction of the 1307 m long and up to 46 m high Pensacola Dam in Oklahoma, USA, from 1938 to 1940.[145]

4.12.1. Deadload

The deadload of one compartment consists of the proper buttress and the adjoining vault

$$G = G_B + G_V \tag{95}$$

4.12.2. Hydrostatic pressure

If the distance between buttresses is denoted by *a*, the components of hydrostatic pressure (see Section 4.10.2) in the horizontal direction are

$$W_h = \gamma_w a\left(\frac{(H-f)^2}{2}\right) \tag{96}$$

and for the upstream slope of the water face of 1 in *n* in the vertical direction are

$$W_v = nW_h \tag{97}$$

4.12.3. Uplift

As in other buttress dams the uplift acts only in the region of the foundation socle below the vault with horizontal projection d. Assuming again a linear distribution of uplift from its maximum value of $m\gamma_w H$ at the upstream face to zero at the end of the longitudinal foundation socle below the vault, the total uplift is given by

$$U = m\gamma_w a\left(\frac{Hd}{2}\right) \tag{98}$$

4.12.4. Adhesion

If A_f denotes the area in plan of the dam's foundation socle of one compartment, the adhesion of the dam to the bedrock is given by

$$C = cA_f \tag{99}$$

4.12.5. Stresses at foundation level

See the theory explained for other buttress dams in Section 4.10.5.

4.12.6. Shrinkage and temperature stresses

See the theory explained for other buttress dams in Section 4.10.6.

4.12.7. Safety against sliding

See the theory explained for other buttress dams in Section 4.10.7.

4.12.8. Crest displacement

See the theory explained for other buttress dams in Section 4.10.8.

4.12.9. Vaults

The durability of multiple arch dams is considerably influenced by cracks in the vault due to the different loadings. Thus, the analysis of vaults must be based on the theory of elasticity[146] and not on the theory of plasticity,[147] although only the latter can yield information on the actual load-carrying capacity of vaults.

4.12.9.1. Deadload

Only the normal component of the vault's deadload

$$g_n = \gamma_c t \sin \beta \quad (100a)$$

causes direct forces and bending moments in the individual arches, whereas the tangential component of

$$g_t = \gamma_c t \cos \beta \quad (100b)$$

in the direction of the vault's generatrix causes only direct forces as long as the fixing moment of the vault to the foundation (Section 5.2) is not considered.

For vaults in the shape of semicircular fixed arches the direct forces at the crown and at the abutment can be approximated by

$$N_y^C = -g_n\left(R - 0{\cdot}727\frac{t^2}{R}\right) \quad (101a)$$

$$N_y^A = -g_n R \quad (101b)$$

The corresponding bending moments approximate to

$$M_y^C = \quad 0{\cdot}242\; g_n t^2 \quad (102a)$$

$$M_y^A = -0{\cdot}484\; g_n t^2 \quad (102b)$$

4.12.9.2. Hydrostatic pressure

The hydrostatic pressure at a depth x below the water level always acts normal to the water face of the vault

$$p = \gamma_w x \quad (103)$$

Then, the relevant interior forces of a semicircular arch are approximated by

$$N_y^{\mathrm{C}} = -p\left(R + \frac{t}{2} - 0{\cdot}727\frac{t^2}{R}\right) \tag{104a}$$

$$N_y^{\mathrm{A}} = -p\left(R + \frac{t}{2}\right) \tag{104b}$$

$$M_y^{\mathrm{C}} = 0{\cdot}242\ pt^2 \tag{105a}$$

$$M_y^{\mathrm{A}} = -0{\cdot}484\ pt^2 \tag{105b}$$

4.12.9.3. *Shrinkage of concrete*

As the concrete strain equals the specific shrinkage ε_s, the interior forces of a semicircular arch are approximately given by

$$N_y^{\mathrm{C}} = -0{\cdot}727\frac{E_c\varepsilon_s t^3}{R^2} \tag{106a}$$

$$N_y^{\mathrm{A}} = 0 \tag{106b}$$

$$M_y^{\mathrm{C}} = -0{\cdot}242\frac{E_c\varepsilon_s t^3}{R} \tag{107a}$$

$$M_y^{\mathrm{A}} = -0{\cdot}484\frac{E_c\varepsilon_s t^3}{R} \tag{107b}$$

4.12.9.4. *Variations of temperature*

Equations (106) and (107) remain valid if the specific shrinkage ε_s is replaced by the temperature strain ωT.

4.12.9.5. *Stresses*

In a semicircular arch of constant thickness with the width one, the stresses can be computed with the section area $A = t$ and the section modulus $S = t^2/6$.

4.13. Numerical example: Gleno Dam

The construction of this multiple arch dam[93,144] began under unfavourable circumstances. The original construction permit had been issued for a massive gravity dam with a slight curvature in plan. The torrent's gorge eroded by the run-off of a glacier was closed with a plug of ashlar masonry (height 23 m, crest width 27 m and crest length 80 m). At the bottom of the gorge an outlet was installed. To save the private client's money the original design was replaced by a multiple arch dam (Fig. 77) and a new construction permit was requested. As the construction started immediately after this decision had been taken, the dam was finished before the second construction permit had been issued. Thus, the client saved half the cost of the original dam (at that time 10 million liras). After heavy rains had filled the reservoir with 5×10^6 m^3 of water and the spillway had started operation, the Gleno Dam collapsed on 1 December 1923 by sliding upon the ashlar masonry plug and partly on the smooth bedrock.

4.13.1. *Deadload*

The deadload of one dam compartment (Fig. 77) consisted of three portions:

(a) buttress socle

$$G_S = 0{\cdot}0215 \times 2{\cdot}00 \times 4{\cdot}44 \times 30{\cdot}00 = 5{\cdot}73 \text{ MN}$$

(b) buttress

$$G_B = 0{\cdot}0215\left(\frac{29{\cdot}50}{2}\right)(2{\cdot}72 \times 26{\cdot}85 + 2{\cdot}00 \times 1{\cdot}80)$$
$$= 24{\cdot}30 \text{ MN}$$

(c) vault

$$G_V = 0{\cdot}0215 \times \frac{\pi}{2}(0{\cdot}80 \times 3{\cdot}40 + 0{\cdot}30 \times 3{\cdot}15)\left(\frac{29{\cdot}50}{0{\cdot}800}\right)$$
$$+ 0{\cdot}0215 \times 6{\cdot}00 \times 4{\cdot}50 \times 3{\cdot}00$$
$$= 4{\cdot}57 + 1{\cdot}74 = 6{\cdot}31 \text{ MN}$$

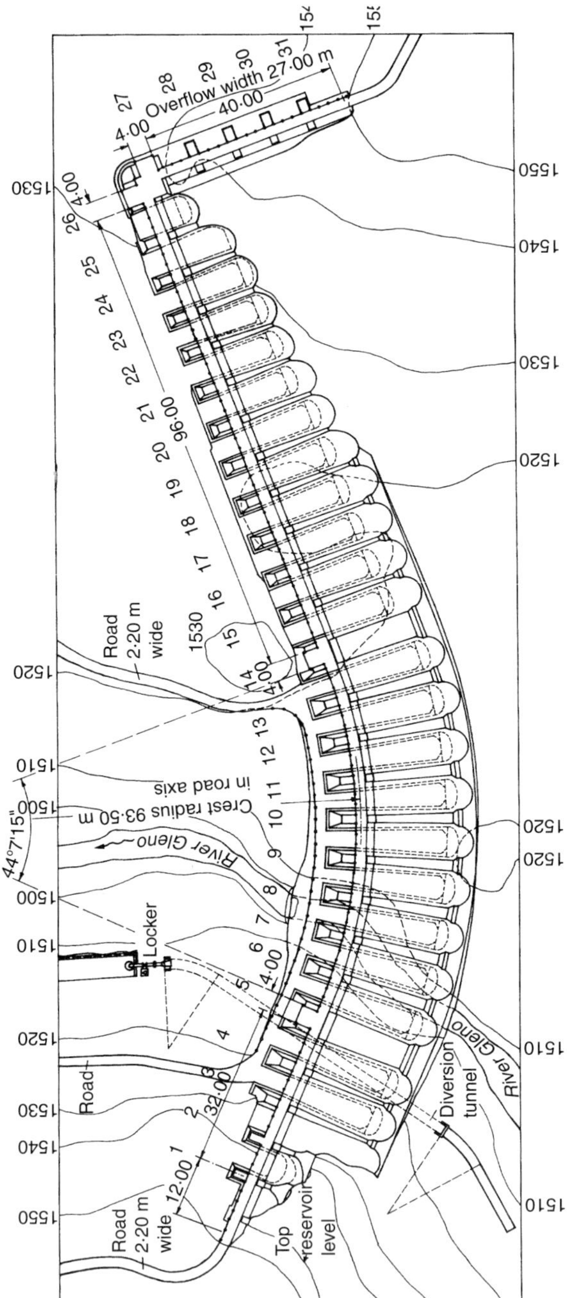

Fig. 77 (above and facing). Gleno multiple arch dam in Italy: (a) plan, (b) cross-section and (c) view[144]

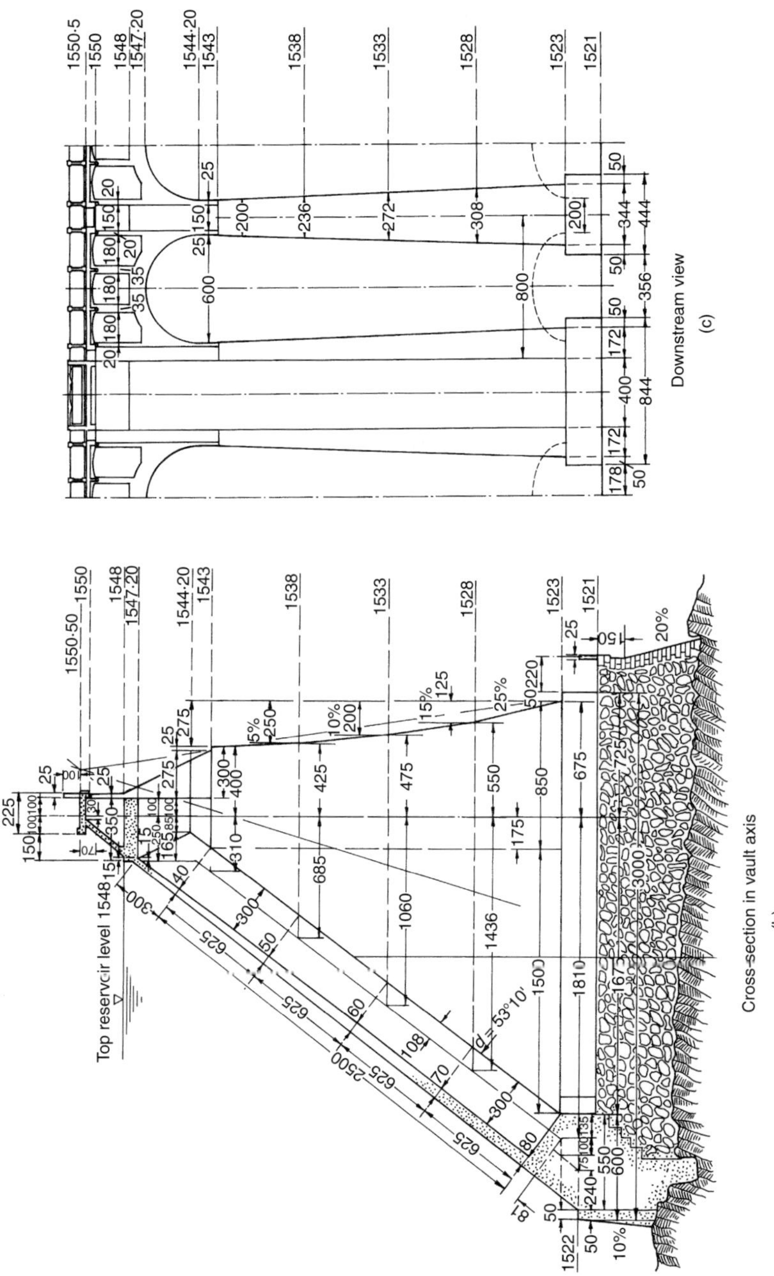

Fig. 77. Continued

In total

$$G = 5{\cdot}73 + 24{\cdot}30 + 6{\cdot}31 = 36{\cdot}34 \text{ MN}$$

were acting on top of the ashlar masonry plug at El. 1521 m.

4.13.2. *Hydrostatic pressure*

According to Equations (96) and (97) the two components of hydrostatic pressure on one compartment amounted to

$$W_h = 0{\cdot}01 \times 8{\cdot}00\left(\frac{27{\cdot}00^2}{2}\right) = 29{\cdot}16 \text{ MN}$$

$$W_v = 0{\cdot}800 \times 29{\cdot}16 = 23{\cdot}33 \text{ MN}$$

4.13.3. *Uplift*

With an uplift ratio of $m = 1{\cdot}0$ the uplift acting on one compartment of the longitudinal foundation socle below the vault is given by

$$U = 1{\cdot}0 \times 0{\cdot}01 \times 27{\cdot}00\left(\frac{8{\cdot}00 \times 6{\cdot}00}{2}\right) = 6{\cdot}48 \text{ MN}$$

4.13.4. *Adhesion*

Because of the low masonry strength and the complete saturation of the mortar with water, no adhesion of the dam upon its underground (ashlar masonry plug and bedrock) is taken into account.

4.13.5. *Stresses at foundation level*

With the section values of the foundation socle $A = 154{\cdot}6 \text{ m}^2$, $S_u = 952 \text{ m}^3$ and $S_d = 763 \text{ m}^3$, and with the interior forces given by Equations (86) and (87) of

$$N = -36{\cdot}34 - 23{\cdot}33 + 6{\cdot}48 = -53{\cdot}19 \text{ MN}$$

$$\begin{aligned} M &= 29{\cdot}16\left(\frac{27{\cdot}00}{3}\right) + 36{\cdot}34 \times 0{\cdot}95 - 23{\cdot}33 \\ &\quad \times 6{\cdot}14 + 6{\cdot}48 \times 11{\cdot}34 \\ &= 262{\cdot}4 + 34{\cdot}5 - 143{\cdot}2 + 73{\cdot}5 = 227{\cdot}2 \text{ MN/m} \end{aligned}$$

the vertical stresses at the dam faces given by Equations (88) amount to

$$\sigma_u = -\frac{53{\cdot}19}{154{\cdot}6} + \frac{227{\cdot}2}{952} = -0{\cdot}34 + 0{\cdot}24 = -0{\cdot}10\ \text{MN/m}^2$$

$$\sigma_d = -\frac{53{\cdot}19}{154{\cdot}6} - \frac{227{\cdot}2}{763} = -0{\cdot}34 - 0{\cdot}30 = -0{\cdot}64\ \text{MN/m}^2$$

4.13.6. Stresses at El. 1525 m

With the section values $A = 66{\cdot}5\ \text{m}^2$, $S_u = 279\ \text{m}^3$ and $S_d = 306\ \text{m}^3$, and the interior forces after Equations (86) and (87) of

$$N = -27{\cdot}75 - 16{\cdot}93 = -44{\cdot}68\ \text{MN}$$

$$M = 21{\cdot}16\left(\frac{23{\cdot}00}{3}\right) + 27{\cdot}75 \times 2{\cdot}55 - 16{\cdot}93 \times 5{\cdot}07$$

$$= 162{\cdot}2 + 70{\cdot}8 - 85{\cdot}8 = 147{\cdot}2\ \text{MN/m}$$

the vertical stresses at the dam faces given by Equations (88) amount to

$$\sigma_u = -\frac{44{\cdot}68}{66{\cdot}5} + \frac{147{\cdot}2}{279} = -0{\cdot}67 + 0{\cdot}53 = -0{\cdot}14\ \text{MN/m}^2$$

$$\sigma_d = -\frac{44{\cdot}68}{66{\cdot}5} - \frac{147{\cdot}2}{306} = -0{\cdot}67 - 0{\cdot}47 = -1{\cdot}14\ \text{MN/m}^2$$

From Equation (90) the shear stress is

$$\tau = \frac{21{\cdot}16}{2{\cdot}00 \times 25{\cdot}75} = 0{\cdot}41\ \text{MN/m}^2$$

The stresses of the inclined dam faces are given by Equations (89) as

$$\max\ \sigma_u = -\frac{0{\cdot}14}{0{\cdot}781} = -0{\cdot}18\ \text{MN/m}^2$$

$$\max\ \sigma_d = -\frac{1{\cdot}14}{0{\cdot}970} = -1{\cdot}18\ \text{MN/m}^2$$

4.13.7. Shrinkage and temperature stresses

The horizontal tensile stress in the buttress of ashlar masonry with concrete facings are about (Equation (91))

$$\sigma_{\mathrm{h}} = \left(5 + 0 + \frac{19}{2}\right) \times 10^{-5} \times 10\,000 = 1{\cdot}45\,\mathrm{MN/m^2}$$

4.13.8. Principal stresses

The vertical stress $\sigma_x = -0{\cdot}67\ \mathrm{MN/m^2}$, the horizontal stress $\sigma_y = 1{\cdot}45\ \mathrm{MN/m^2}$ and the shear stress $\tau_{xy} = 0{\cdot}41\ \mathrm{MN/m^2}$ result in principal stresses in the centre-line of the buttress' section at El. 1525 m of (Equation (93))

$$\sigma_{1,2} = \frac{-0{\cdot}67 + 1{\cdot}45}{2} \pm \left[\left(\frac{-0{\cdot}67 - 1{\cdot}45}{2}\right)^2 + 0{\cdot}41^2\right]^{1/2}$$

$$= 0{\cdot}39 \pm (1{\cdot}12 + 0{\cdot}17)^{1/2} = \begin{matrix} 1{\cdot}53 \\ -0{\cdot}75 \end{matrix}\ \mathrm{MN/m^2}$$

Deviation of the principal compressive stresses from the vertical is given by Equation (94):

$$\tan\ 2\alpha = \frac{2 \times 0{\cdot}41}{-0{\cdot}67 - 1{\cdot}45} = -0{\cdot}3868$$

and $\alpha = 10^\circ\ 34'$ respectively. As the principal tensile stress surpassed the tensile strength of the low-quality ashlar masonry, cracks developed in the buttresses.[93,144]

4.13.9. Safety against sliding

From the actual sliding of the buttress socle upon the ashlar masonry plug it follows, according to Equation (92), that the coefficient of friction at foundation level (disregarding adhesion) amounted to

$$\tan\ \phi = \frac{29{\cdot}16}{36{\cdot}34 + 23{\cdot}33 - 6{\cdot}48} = 0{\cdot}548$$

The corresponding angle of friction of $\phi = 28^\circ\ 43'$ is highly probable.

4.13.10. Vault

Only the cross-section at the vault's base is considered, disregarding the actual fixation in the direction of the vault's generatrix.

4.13.10.1. Deadload

The following values are determined from the equations listed.

Equation (100*a*) $g_n = 0{\cdot}0215 \times 0{\cdot}80 \times 0{\cdot}5995 = 0{\cdot}010\ \text{MN/m}^2$

Equation (101*b*) $N_y^A = -0{\cdot}010 \times 3{\cdot}40 = -0{\cdot}034\ \text{MN/m}$

Equation (102*b*) $M_y^A = -0{\cdot}484 \times 0{\cdot}010 \times 0{\cdot}80^2$

$= -0{\cdot}0031\ \text{MN m/m}$

4.13.10.2. Hydrostatic pressure

Equation (103) $p = 0{\cdot}01 \times 23{\cdot}00 = 0{\cdot}230\ \text{MN/m}^2$

Equation (104*b*) $N_y^A = -0{\cdot}23 \times 3{\cdot}80 = -0{\cdot}874\ \text{MN/m}$

Equation (105*b*) $M_y^A = -0{\cdot}484 \times 0{\cdot}23 \times 0{\cdot}80^2$

$= -0{\cdot}0713\ \text{MN m/m}$

4.13.10.3. Shrinkage of concrete

Equation (106*b*) $N_y^A = 0$

Equation (107*b*) $M_y^A = -0{\cdot}484 \times 10\,000 \times 5 \times 10^{-5}\left(\frac{0{\cdot}80^3}{3{\cdot}40}\right)$

$= -0{\cdot}0365\ \text{MN m/m}$

4.13.10.4. Variation of temperature

Equation (1) $T = \frac{62}{0{\cdot}80 + 2{\cdot}4} = \pm 19\ ^\circ\text{C}$

For the time of year of the dam's collapse (1 December) only half this value is accounted for.

Equation (106*b*) $N_y^A = 0$

Equation (107*b*) $M_y^A = -0{\cdot}484 \times 10\,000\left(\frac{19}{2}\right)10^{-5}\left(\frac{0{\cdot}80^3}{3{\cdot}40}\right)$

$= -0{\cdot}0693\ \text{MN m/m}$

4.13.10.5. Superposition

Only the critical abutment section is considered.

$$N_y^A = -0{\cdot}034 - 0{\cdot}874 - 0 - 0 = -0{\cdot}908 \text{ MN/m}$$

$$M_y^A = -0{\cdot}0031 - 0{\cdot}0713 - 0{\cdot}0365 - 0{\cdot}0693$$
$$= -0{\cdot}1802 \text{ MN m/m}$$

4.13.10.6. Stresses

$$\sigma_y^A = -\frac{0{\cdot}908}{0{\cdot}80} \pm \frac{6 \times 0{\cdot}1802}{0{\cdot}80^2}$$
$$= -1{\cdot}14 \pm 1{\cdot}69 = \begin{matrix} 0{\cdot}55 \\ -2{\cdot}83 \end{matrix} \text{ MN/m}^2$$

5. Arch dams

5.1. Dam types

Designers of even the oldest masonry dams tried intuitively to enhance the resistance to hydrostatic pressure by giving them a curvature in plan. With *Mariotte*'s (1620 to 1686) ring formula

$$\sigma = pR/t \tag{108}$$

it became possible to account for this curvature.[148] The first modern arch dam was built after the design by *F. Zola* — the father of the well known French writer *E. Zola* — for the water supply of Aix-en-Provence[149] (height $H = 37{\cdot}7$, crest length $B = 63{\cdot}5$ m, base thickness $t_u = 12{\cdot}75$ m, exterior radius of curvature $R_e = 48{\cdot}2$ m, volume of ashlar masonry with cut-stone facings $V = 10\ 100$ m^3) from 1843 to 1852. This dam has already served its purpose for 147 years without any problem besides sedimentation. The next step in the development of arch dams[150] was the daring Bear Valley Dam by *Brown* ($H = 18{\cdot}9$ m, $B = 75{\cdot}3$ m, $t_u = 6{\cdot}75$ m, $R_e = 91{\cdot}5$ m) on the young Santa Ana River in the San Bernardino Mountains of California. From its description in a then well-known American book on dams[130] the Bear Valley Dam initiated the construction of arch dams all over the world. Towards the end of the 19th Century a whole series of arch dams with sparing dimensions[151] had been built in New South Wales, Australia. Even a century later these dams are still remarkable (Table 12) because of the achieved ratio of crest length to height (e.g. Parkes Dams in 1897, $H = 10{\cdot}2$ m, $B = 164{\cdot}6$ m, $B/H = 16{\cdot}1$).

All arch dams built up to then possessed a constant curvature in plan and were therefore called 'constant radius arch dams'. For better utilization of the construction material (masonry or concrete), in 1910 *Eastwood*[47] proposed to diminish the radius of curvature from the dam's crest to its base. By this concept the angle of aperture of dif-

Table 12. Older constant radius dams in Australia

Completion	Name	H: m	B: m	B/H
1896	Lithgow I	10·7	54·3	5·1
1897	Parkes	10·2	164·6	16·1
1898	Cootamundra	14·0	195·0	13·9
	Parramatta	16·0	68·6	4·3
	Tamworth	18·6	134·1	7·2
1899	Wellington	14·6	106·7	7·3
	Mudgee	15·2	151·8	10·0
	Wollongong	12·8	163·1	12·7

ferent horizontal sections remains of the same magnitude. From 1913 to 1914 *Jorgensen*[46] built the first 'constant angle arch dam' on the Salmon Creek in Alaska.

Milestones in the development of arch dams are given in Tables 13[150] and Tables 14 – 18 list maximum characteristics of arch dams.[150]

5.2. Approximate analysis using shell theory

Several trials[33,58,59] were undertaken to take advantage of tank theory in the analysis of arch dams. But it was not until 1992 that *Herzog*[64,152] succeeded in showing that even the most radical simplification of the problem to the use of closed formulae for circular tanks of constant thickness[153] suffices for practical purposes as adequate agreement between predicted and observed behaviour of arch dams (displacements and strains or stresses) can be achieved.

The approximations consist firstly of the assumption of a rigidly fixed dam base, whereas real arch dams are at best elastically fixed in the bedrock. Secondly, bending in the horizontal direction is considered only for midheight, and this with an empirical formula. Although its theoretical derivation was not achieved up to now, the prediction of displacements and strains coincides astonishingly well with a large number of arch dams of very different shapes and sizes.[64]

5.2.1. Deadload

Arch dams are like massive gravity dams constructed as individual concrete blocks (square columns with sectional side lengths hardly surpassing 15 m to 20 m) to counteract the formation of shrinkage and cooling cracks. The deadload of arch dams is generally conveyed

Table 13 (pages 129 and 130). Milestones of arch dam construction

No	Name of dam	Location	Construction period	H: m	B: m	V: 10^3 m^3	t_0: m	t_u: m	B/H	Dam type[a]	Bedrock
1	Zola	France	1843 – 1852	38	64	10·1	5·80	12·75	1·68	CR	limestone
2	Bear Valley	USA	1883 – 1884	18	75	2·5	0·97	6·75	4·12	CR	granite
3	Sweetwater	USA	1886 – 1888	28	104	16·0	3·65	13·7	3·78	CR	
4	Parkes	Australia	1896 – 1897	10	165		0·91	4·10	16·1	CR	granite
5	Cootamundra	Australia	1897 – 1898	14	195		0·91	3·96	13·9	CR	granite
6	Cheesman	USA	1900 – 1904	72	216	76·5	5·5	54	3·00	CR	granite
7	Six Mile Creek	USA	1902 – 1903	27	30	1·0	0·45	2·4	1·11	Cup	schist
8	Buffalo Bill	USA	1905 – 1909	99	61	61·4	3·05	32·9	0·62	CR	granite
9	Roosevelt	USA	1906 – 1911	85	221	273	4·9	56·2	2·60	CR	sandstone
10	Salmon Creek	USA	1913 – 1914	52	166	38·5	1·83	14·6	3·20	CA	conglomerate
11	Arrowrock	USA	1912 – 1915	108	351	434	4·9	68	3·25	CR	granite
12	Montsalvens	Switzerland	1919 – 1920	55	110	26	2·00	22·5	2·00	CA	limestone
13	Gibraltar	USA	1919 – 1920	56	300	44·5	2·44	19·8	5·32	CR	
14	Montejaque	Spain	1922 – 1924	83	78	29	3·0	16·8	0·94	CA	limestone
15	Stevenson Creek	USA	1926	18	43	0·37	0·61	2·29	2·33	CR	granite
16	Pacoima	USA	1926 – 1929	113	195	117	3·2	30·5	1·72	CA	diorite
17	Diablo	USA	1927 – 1930	118	360	260	4·9	44·5	3·05	CA	granite
18	Owyhee	USA	1928 – 1932	127	254	387	9·2	81	2·00	CR	lava
19	Hoover	USA	1932 – 1936	221	372	2490	13·7	201	1·69	CR	andesite, tuff breccia
20	Marèges	France	1934 – 1936	90	248	183	3·0	21·0	2·69	CA	granite
21	Osiglietta	Italy	1937 – 1939	76	224	75	3·6	10·7	2·95	Cup	limestone
22	Lumiei	Italy	1942 – 1947	136	138	104	3·5	13·7	1·01	CA	dolomite
23	Ross	USA	1943 – 1949	165	397	672	10·0	63·4	2·41	CA	granitic gneiss
24	Pieve di Cadore	Italy	1946 – 1949	63	406	377	6·0	26·2	6·44	CR	dolomite

Table 13. Continued

No	Name of dam	Location	Construction period	H: m	B: m	V: 10^3 m^3	t_0: m	t_u: m	B/H	Dam type[a]	Bedrock
25	Hungry Horse	USA	1948 – 1952	172	645	2200	10·7	103·5	3·75	CR	
26	Tignes	France	1949 – 1953	180	296	632	10·0	44·6	1·63	CR	quartzite
27	Malpasset	France	1953 – 1955	66	245	48	1·50	6·9	3·71	CA	gneiss
28	Mauvoisin	Switzerland	1951 – 1957	237	520	2030	14·0	53·5	2·19	CA	mica schist
29	Kariba	Zimbabwe	1956 – 1959	125	615	975	13·0	18·5	4·92	CA	gneiss
30	Vaiont	Italy	1957 – 1960	261	190	352	3·4	22·7	0·73	CA	limestone
31	Valle di Lei	Switzerland	1958 – 1961	143	690	850	16·5	28·0	4·82	Cup	schist
32	Schiffenen	Switzerland	1961 – 1963	47	417	185	7·0	13·5	8·88	CR	molasse
33	Toules	Switzerland	1960 – 1963	84	460	255	4·5	17·9	5·35	Cup	paragneiss
34	Glen Canyon	USA	1960 – 1964	217	476	3750	10·7	91·5	2·20	CR	sandstone
35	Mossyrock	USA	1965 – 1968	185	530	973	8·2	39·8	2·87	CA	basalt, andesite
36	Verwoerd	South Africa	1966 – 1970	90	948	1800			10·53	CR	dolerite
37	Kölnbrein	Austria	1973 – 1977	200	626	1580	7·6	37·0	3·13	CA	gneiss
38	Inguri	Russia	1964 – 1981	271	760	3960	10·0	86	2·80	CA	limestone
39	Sayanoshushensk	Siberia	1970 – 1988	245	1066	9120	25·0	105·7	4·35	CR	cristalline schist

[a]CR = constant radius dam, CA = constant angle dam, Cup = cupola dam

Table 14. Highest constant radius arch dams

Completion date	Name	Location	Height: m
1852	Zola	France	37·7
1904	Cheesman	USA	72
1909	Buffalo Bill	USA	99
1915	Arrowrock	USA	108
1932	Owyhee	USA	127
1936	Hoover	USA	221
1988	Sayanoshushensk	Siberia	245

Table 15. Highest constant angle arch dams

Completion date	Name	Location	Height: m
1914	Salmon Creek	USA	52
1920	Montsalvens	Switzerland	55
1924	Montejaque	Spain	83
1929	Pacoima	USA	113
1930	Diablo	USA	118
1947	Lumiei	Italy	136
1949	Ross	USA	165
1957	Mauvoisin	Switzerland	237
1960	Vaiont	Italy	261
1981	Inguri	Russia	271

to the bedrock in the vertical direction (Fig. 78)

$$N_x = -G \tag{109}$$

Cupola-type arch dams with a pronounced vertical curvature need props in the case of an upstream overhang (1935 to 1936 Marèges Dam in the French Massif Central[154]). In the case of a downstream overhang the vertical construction (block) joints must be closed early enough to warrant the dam's ability to activate compressive ring stresses (1937 to 1939 Osiglietta Dam in Italy[155]).

5.2.2. Hydrostatic pressure

With the help of the characteristic length of a circular cylindrical shell for the dam's base (Fig, 79) of

$$\lambda_x = 0{\cdot}76\sqrt{R_{yu}t_u} \tag{110}$$

Table 16. Longest crests of arch dams

Completion date	Name	Location	Length: m
1852	Zola	France	63·5
1884	Bear Valley	USA	75
1888	Sweetwater	USA	104
1897	Parkes	Australia	165
1898	Cootamundra	Australia	195
1904	Cheesman	USA	216
1911	Roosevelt	USA	221
1915	Arrowrock	USA	351
1930	Diablo	USA	360
1936	Hoover	USA	372
1949	Ross	USA	397
	Pieve di Cadore	Italy	406
1952	Hungry Horse	USA	645
1961	Valle di Lei	Switzerland	690
1970	Verwoerd	South Africa	948
1988	Sayanoshushensk	Siberia	1066

Table 17 Largest ratios of crest length to height

Completion date	Name	Location	B/H	Height: m
1852	Zola	France	1·68	37·7
1884	Bear Valley	USA	4·12	18·3[a]
1897	Parkes	Australia	16·1	10·2[a]
1920	Gibraltar	USA	5·95	56
1931	Nattai Creek	Australia	16·8	7·65[a]
1949	Pieve di Cadore	Italy	6·44	63
1963	Schiffenen	Switzerland	8·88	47
1970	Verwoerd	South Africa	10·53	90

[a]Dams below 30 m height are not considered as large dams by ICOLD (International Commission on Large Dams)

the closed formulae run for the fixing moment as

$$M_x^{\mathrm{W}} = \gamma_{\mathrm{w}}(H - \lambda_x)\frac{\lambda_x^2}{2} \tag{111}$$

and for the shear force as

$$Q_x^{\mathrm{W}} = \gamma_{\mathrm{w}}\lambda_x\left(H - \frac{\lambda_x}{2}\right) \tag{112}$$

Table 18. Largest concrete volumes of arch dams

Completion data	Name	Location	Volume: 10^3 m^3
1852	Zola	France	10·1
1888	Sweetwater	USA	16·0
1904	Cheesman	USA	76·5
1911	Roosevelt	USA	273
1915	Arrowrock	USA	434
1936	Hoover	USA	2490
1964	Glen Canyon	USA	3750
1981	Inguri	Russia	3960
1988	Sayanoshushensk	Siberia	9075

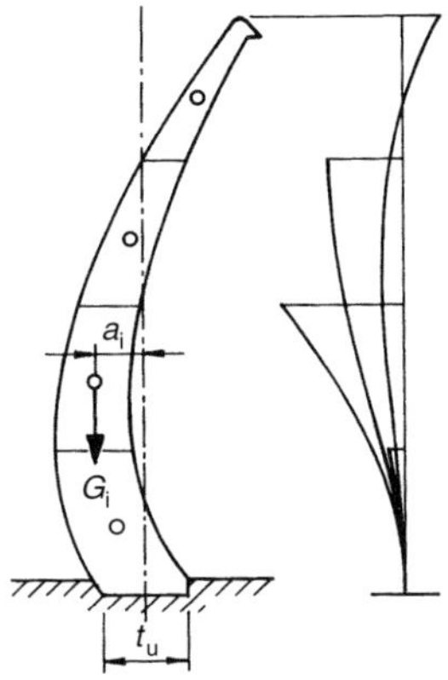

Fig. 78. Deformations of an individual block of an arch dam during construction[136]

Both interior forces die away in the vertical direction like damped oscillations (Fig. 80).

The remaining portion of the water load

$$W_y = \gamma_W\left(\frac{H^2}{2}\right) - Q_x^W = \gamma_W H\left[\frac{H}{2} - \frac{\lambda_x}{H}\left(H - \frac{\lambda_x}{2}\right)\right]$$
$$= \gamma_W H\left(\frac{H}{2} - \lambda_x + \frac{\lambda_x^2}{2H}\right) \qquad (113)$$

must be conveyed to the valley flanks in the horizontal direction. The ratio of its average value

$$p_y = \frac{W_y}{H} = \gamma_w \left(\frac{H}{2} - \lambda_x + \frac{\lambda_x^2}{2H} \right) \tag{114}$$

to the hydrostatic pressure at midheight is a transparent measure of

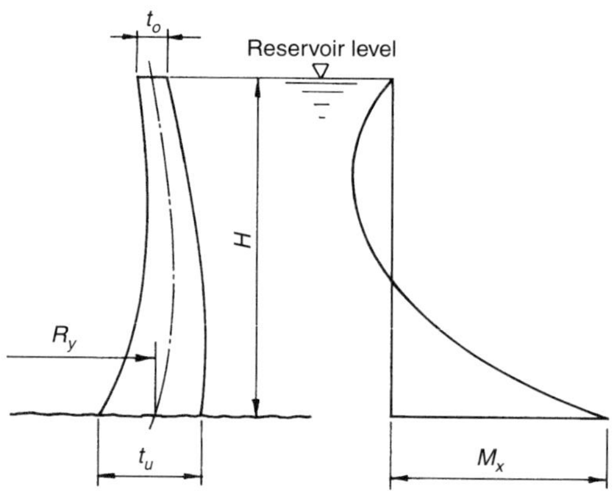

Fig. 79. Vertical bending of an arch dam[64]

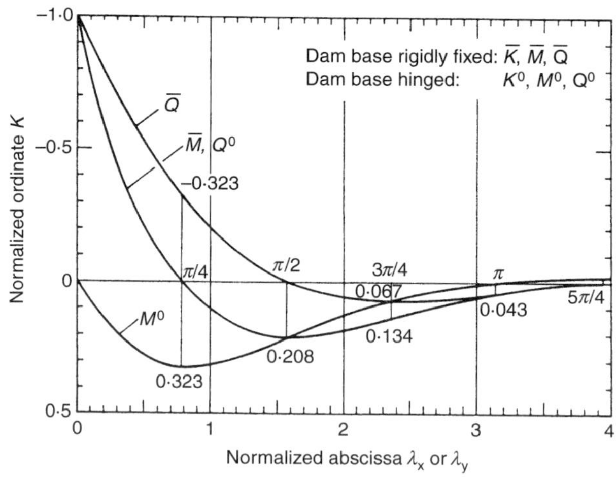

Fig. 80. Dying away of bending moments and shear forces from the fixed base of a circular cylindrical shell in meridional direction[64]

the efficiency

$$\eta = 2p_y/\gamma_w H \tag{115}$$

of the dam's arch action.

The characteristic length of a circular cylindrical shell at midheight

$$\lambda_y = 0{\cdot}76\sqrt{R_{ym} t_m} \tag{116}$$

gives the horizontal fixing moment in the abutment at midheight according to

$$M_{yA}^{W} = p_y \frac{\lambda_y^2}{2} \tag{117}$$

The adequacy of this empirical formula has been checked by the posterior analysis of some 70 arch dams. This fixing moment in the abutment also dies away in the horizontal direction (Fig. 81) like a

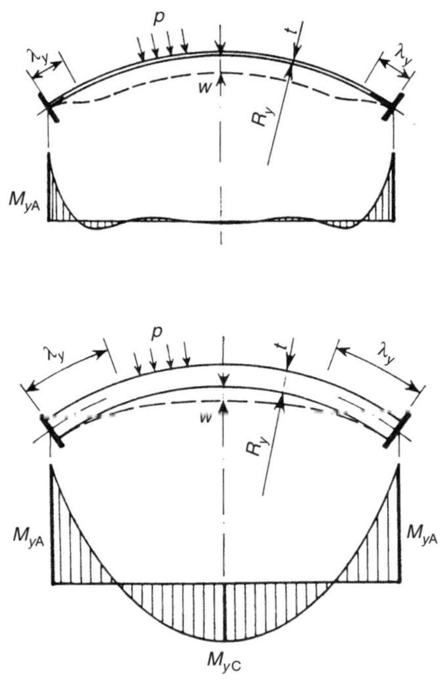

Fig. 81. Influence of an arch dam's wall thickness on the dying away of bending moments from the abutment in the horizontal ring direction[64]

damped oscillation. The direct force in the horizontal direction at midheight due to hydrostatic pressure amounts (after *Mariotte*) to

$$N_y^{\mathrm{W}} = p_y\left(R_{y\mathrm{m}} + \frac{t_\mathrm{m}}{2}\right) \tag{118}$$

5.2.3. Variation of temperature

The fixing moments of an arch dam due to uniform variations of temperature after Equation (1) are approximated by

$$M_x^{\mathrm{T}} = \frac{E_\mathrm{c}\omega T t_\mathrm{u}}{R_{y\mathrm{u}}} \times \frac{\lambda_x^2}{2} \tag{119}$$

for the dam's base and by

$$M_{y\mathrm{A}}^{\mathrm{T}} = \frac{E_\mathrm{c}\omega T t_\mathrm{m}}{R_{y\mathrm{m}}} \times \frac{\lambda_y^2}{2} \tag{120}$$

for the abutments, while the ring force at midheight amounts to

$$N_{y\mathrm{m}}^{\mathrm{T}} = E_\mathrm{c}\omega T t_\mathrm{m} \tag{121}$$

5.2.4. Shrinkage of concrete

If the temperature strain ωT is replaced by the specific shrinkage of concrete ε_s, the formulae for a decrease of temperature are also valid for concrete shrinkage.

5.2.5. Stresses

With the section area $A = t$ and the section modulus $S = t^2/6$ the vertical stresses at the dam's base amount to

$$\sigma_x = \frac{N_x}{t_\mathrm{u}} \pm \frac{6M_x}{t_\mathrm{u}^2} \tag{122}$$

and the horizontal stresses at midheight to

$$\sigma_y = \frac{N_y}{t_\mathrm{m}} \pm \frac{6M_y}{t_\mathrm{m}^2} \tag{123}$$

The vertical stress in the centre-line of the base section is

$$\sigma_{x0} = \frac{N_x}{t_u} \tag{124}$$

and the maximum shear stress (parabolic distribution) is

$$\tau_x = \frac{3}{2} \times \frac{Q_x}{t_u} \tag{125}$$

These give the principal stresses according to the equation

$$\sigma_{1,2} = \frac{\sigma_{x0}}{2} \pm \left[\left(\frac{\sigma_{x0}}{2} \right)^2 + \tau_x^2 \right]^{1/2} \tag{126}$$

and the deviation of the principal compressive stress from the vertical

$$\tan\ 2\alpha = \frac{2\tau_x}{\sigma_{x0}} \tag{127}$$

If the principal tensile stress exceeds the tensile strength of concrete, cracks and all their possible consequences must be expected in the dam's interior.[156]

5.2.6. Radial displacements

The radial displacement of an arch dam at midheight consists of two portions. The portion due to the normal force is in the case of hydrostatic pressure

$$w_{\mathrm{mN}}^{\mathrm{W}} = \frac{N_y^{\mathrm{W}} R_{y\mathrm{m}}}{E_{\mathrm{c}} t_{\mathrm{m}}} \tag{128}$$

and in the case of a temperature variation

$$w_{\mathrm{mN}}^{\mathrm{T}} = \omega T R_{y\mathrm{m}} \tag{129}$$

The portion due to the fixing moment (Fig. 82) is

$$w_{\mathrm{mM}} = \int \frac{M_y^0 M_y'}{E_{\mathrm{c}} I} \mathrm{d}y = \frac{9 M_{y\mathrm{A}} \lambda_y^2}{E_{\mathrm{c}} t_{\mathrm{m}}^3} \tag{130}$$

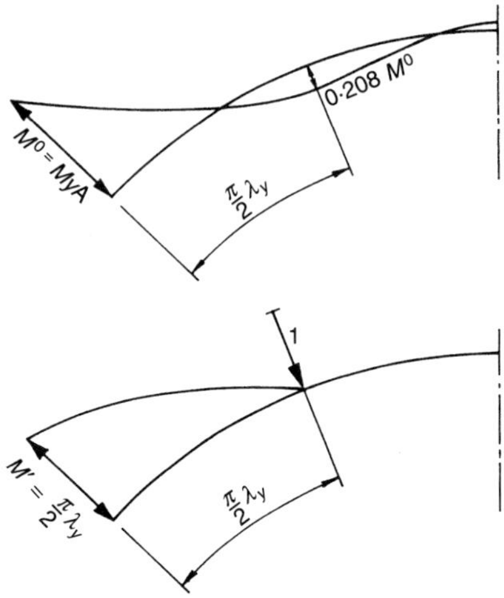

Fig. 82. Portion of radial displacement due to the horizontal bending moment at the abutment (approximation)[64]

5.2.7. Elasticity of bedrock

The influence of bedrock elasticity can, for the ratio of Young's modulus of rock to that of concrete, be considered with the reduction

$$M_{\text{red}} = M\left(\frac{E_{\text{r}}}{E_{\text{c}}}\right) \tag{131}$$

both in vertical and horizontal directions. Due to the reduction of the negative fixing moment at the abutment M_{A}, the positive moment in the span M_i is increased. In the case of a rigid fixation the positive moment amounts to 0·208 M_{A} and in the case of a fictional hinge at the abutment to 0·323 M_{A} (Fig. 80). For unequal moduli of elasticity, i.e. $E_{\text{r}} \neq E_{\text{c}}$, the linear approximation for the positive span moment

$$M_i = 2M_{\text{A}}\left[\bar{K} + (K^0 - \bar{K})\left(1 - \frac{E_{\text{r}}}{E_{\text{c}}}\right)\right] \tag{132}$$

can be used. The normalized ordinates $\bar{K}$ for rigid fixation and K^0 for a fictional hinge at the abutment are given in Fig. 80 as functions of the normalized abscissae x/λ_x and y/λ_y. The factor of 2 in Equation (132) accounts for the fact that the fixing moments die away in the horizontal direction from two abutments.

5.2.8. Upward sliding on slightly inclined valley flanks

This extraordinary possibility of failure was brought to the attention of dam engineers by the collapse of a small constant radius arch dam in China.[97,157] If G denotes the deadload of half the dam and

$$N_y = \int_0^H p_y R_{ye}\, dx = W_y\left(R_{ym} + \frac{t_m}{2}\right) \tag{133}$$

is the sum of all horizontal direct forces due to hydrostatic pressure in the dam's vertical plane of symmetry, the resultant force at the inclined valley flank (Fig. 83) is approximately

$$R = \left[G^2 + \left(N_y \cos \frac{\omega}{2}\right)^2\right]^{1/2} \tag{134}$$

and its inclination is given by

$$\tan \alpha = \frac{G}{N_y \cos \omega/2} \tag{135}$$

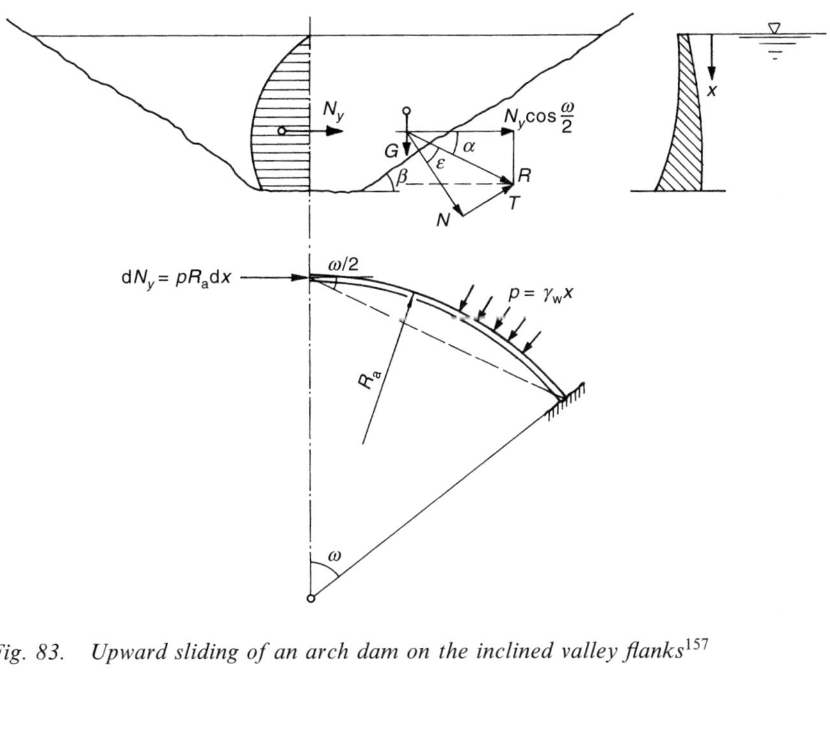

Fig. 83. Upward sliding of an arch dam on the inclined valley flanks[157]

From a comparison of angles, the angle between the resultant force and the inclined valley flank is determined as

$$\varepsilon = \frac{\pi}{2} - (\alpha + \beta) \tag{136}$$

where β denotes the inclination of the valley flank. If the actual angle ε is smaller than the angle of friction ϕ of the dam upon the bedrock, upward sliding can occur.

5.3. Numerical example: Stevenson Creek Test Dam (Fig. 84)

Measurements on existing dams are necessary to check theoretical predictions for arch dams. The first observations of displacements on the Barren Jack Creek Dam in Australia[151] showed the dominant influence of temperature variations. Based on *Noetzli*'s proposal, the American Engineering Foundation built the Stevenson Creek Test Dam near Fresno in California.[52] In 1926 painstaking measurements were taken at night to exclude any influence of radiation by the Sun. The report of these tests[158] is now required reading for all designers of arch dams. A summary[159] is now given.

(a) Variations of temperature are of special importance as they influence not only the magnitude and distribution of stresses but also the displacements and the formation of cracks.
(b) The hydration heat of cement should have dissipated before construction joints are grouted or cooling slots filled with concrete.
(c) Even in thin arch dams a considerable portion of hydrostatic pressure is carried to the bedrock by cantilever action.
(d) Horizontal cracks can appear at the dam's heel due to tensile stresses in the cantilever caused by the fixing moment.
(e) Displacements of arch dams can be predicted realistically with help of the theory of elasticity if the bedrock elasticity as well as the torsional and shear deformations of the dam are accounted for.
(f) The hydrostatic load on horizontal arches can in symmetrical dams be constant from crown to abutment. But in most cases they vary to such a degree that the stress analysis of important arch dams should not be based on such a simplifying assumption.
(g) The trial load method[49] is a suitable means of analysing arches and cantilevers of arch dams of arbitrary shape and size.

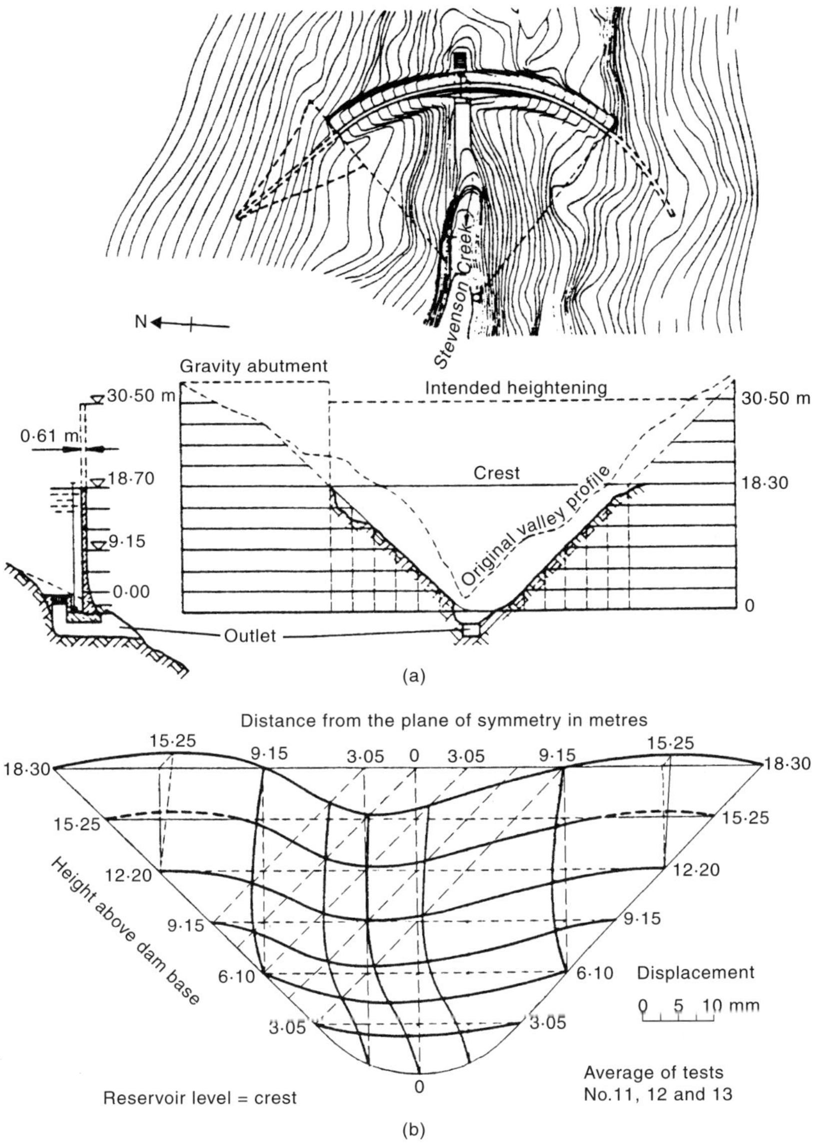

Fig. 84. Stevenson Creek Test Dam[163]

The Stevenson Creek Test Dam had been built to check the predictions of the then new trial load method of the US Bureau of Reclamation. Even today it is used for checks of new methods of analysis.

5.3.1. Approximate analysis

Equation (109) $N_x = -0{\cdot}025\left(0{\cdot}305 \times 18{\cdot}30 + \frac{1{\cdot}983}{3} \times 9{\cdot}15\right)$

$= -0{\cdot}291$ MN/m

Equation (110) $\lambda_x = 0{\cdot}76\sqrt{19{\cdot}66 \times 2{\cdot}29} = 6{\cdot}26$ m

Equation (111) $M_x^{\mathrm{W}} = -0{\cdot}01 \times (18{\cdot}30 - 6{\cdot}26)\left(\frac{6{\cdot}26^2}{2}\right)$

$= -2{\cdot}359$ MN m/m

Equation (112) $Q_x^{\mathrm{W}} = 0{\cdot}01 \times 6{\cdot}26\left(18{\cdot}30 - \frac{6{\cdot}26}{2}\right)$

$= 0{\cdot}950$ MN/m

Equation (114) $p_y = 0{\cdot}01\left(\frac{18{\cdot}30}{2} - 6{\cdot}26 + \frac{6{\cdot}26^2}{2 \times 18{\cdot}30}\right)$

$= 0{\cdot}0396$ MN/m^2

Equation (115) $\eta = \frac{2 \times 0{\cdot}0396}{0{\cdot}01 \times 18{\cdot}30} = 0{\cdot}433$

Equation (116) $\lambda_y = 0{\cdot}76\sqrt{30{\cdot}50 \times 0{\cdot}61} = 3{\cdot}29$ m

Equation (117) $M_{y\mathrm{A}}^{\mathrm{W}} = -0{\cdot}0396\left(\frac{3{\cdot}28^2}{2}\right) = -0{\cdot}213$ MN m/m

Equation (118) $N_y^{\mathrm{W}} = -0{\cdot}0396\left(30{\cdot}50 + \frac{0{\cdot}61}{2}\right)$

$= -1{\cdot}220$ MN/m

$\frac{y}{\lambda_y} = \frac{B_{\mathrm{m}}}{2\lambda_y} = \frac{23{\cdot}52}{2 \times 3{\cdot}28} = 3{\cdot}585$

From Fig. 80, $\bar{\mathrm{K}} = 0{\cdot}014$

$M_{y\mathrm{C}}^{\mathrm{W}} = 2 \times 0{\cdot}014 \times 0{\cdot}213 = 0{\cdot}00596$ MN m/m

Equation (1) $T_x = \frac{62}{2{\cdot}29 + 2{\cdot}4} = \pm 13{\cdot}2\,^{\circ}\mathrm{C}$

Equation (119) $M_x^{\mathrm{T}} = \pm\frac{20\,000 \times 10^{-5} \times 13{\cdot}2 \times 2{\cdot}29}{29{\cdot}66} \times \frac{6{\cdot}26^2}{2}$

$= \pm 3{\cdot}994$ MN m/m

Equation (1) $T_y = \frac{62}{0{\cdot}61 + 2{\cdot}4} = \pm 20{\cdot}6\,^{\circ}\mathrm{C}$

Equation (120) $$M^{\mathrm{T}}_{y\mathrm{A}} = \pm\frac{20\,000 \times 10^{-5} \times 20{\cdot}6 \times 0{\cdot}61}{30{\cdot}50} \times \frac{3{\cdot}28^2}{2}$$

$$= \pm 0{\cdot}443 \text{ MN m/m}$$

Equation (121) $$N^{\mathrm{T}}_{y} = \mp 20\,000 \times 10^{-5} \times 20{\cdot}6 \times 0{\cdot}61$$

$$= \mp 2{\cdot}513 \text{ MN/m}$$

Equation (122) $$\sigma^{\mathrm{G+W}}_{x\mathrm{u}} = -\frac{2 \times 0{\cdot}291}{2{\cdot}29} + \frac{6 \times 2{\cdot}359}{2{\cdot}29^2}$$

$$= -0{\cdot}25 + 2{\cdot}62 = 2{\cdot}37 \text{ MN/m}^2 \text{ (upstream)}$$

$$\sigma^{\mathrm{G+W}}_{x\mathrm{d}} = 0 - 2{\cdot}62$$

$$= -2{\cdot}62 \text{ MN/m}^2 \text{ (downstream)}$$

$$\sigma^{\mathrm{T}}_{x} = \pm\frac{6 \times 3{\cdot}994}{2{\cdot}29^2} = \pm 4{\cdot}57 \text{ MN/m}^2$$

Equation (123) $$\sigma^{\mathrm{W}}_{y\mathrm{A}} = -\frac{1{\cdot}220}{0{\cdot}61} \pm \frac{6 \times 0{\cdot}213}{0{\cdot}61^2} = -2{\cdot}00 \pm 3{\cdot}43$$

$$= \begin{matrix} 1{\cdot}43 \\ -5{\cdot}43 \end{matrix} \text{ MN/m}^2$$

$$\sigma^{\mathrm{W}}_{y\mathrm{C}} = -\frac{1{\cdot}220}{0{\cdot}61} \mp \frac{6 \times 0{\cdot}00596}{0{\cdot}61^2} = -2{\cdot}00 \pm 0{\cdot}10$$

$$= \begin{matrix} -2{\cdot}10 \\ -1{\cdot}90 \end{matrix} \text{ MN/m}^2$$

$$\sigma^{\mathrm{T}}_{y\mathrm{A}} = -\frac{2{\cdot}513}{0{\cdot}61} \mp \frac{6 \times 0{\cdot}443}{0{\cdot}61^2} = -4{\cdot}12 \mp 7{\cdot}14$$

$$= \begin{matrix} -11{\cdot}26 \\ 3{\cdot}02 \end{matrix} \text{ MN/m}^2$$

$$\sigma^{\mathrm{T}}_{y\mathrm{C}} = -\frac{2{\cdot}513}{0{\cdot}61} \pm \frac{6 \times 0{\cdot}01240}{0{\cdot}61^2} = -4{\cdot}12 \pm 0{\cdot}20$$

$$= \begin{matrix} -3{\cdot}92 \\ -4{\cdot}32 \end{matrix} \text{ MN/m}^2$$

Equation (124) $$\sigma_x = -\frac{0{\cdot}291}{2{\cdot}29} = -0{\cdot}13 \text{ MN/m}^2$$

Equation (125) $$\tau_x = \frac{3}{2} \times \frac{0{\cdot}950}{2{\cdot}29} = 0{\cdot}62 \text{ MN/m}^2$$

Equation (126) $$\sigma_{1,2} = -\frac{0{\cdot}13}{2} \pm \left[\left(\frac{0{\cdot}13}{2}\right)^2 + 0{\cdot}62^2\right]^{1/2}$$

$$= -0{\cdot}07 \pm 0{\cdot}63 = \begin{matrix} 0{\cdot}56 \\ -0{\cdot}70 \end{matrix} \ \mathrm{MN/m^2}$$

Equation (127) $\tan 2\alpha = -\dfrac{2 \times 0{\cdot}62}{0{\cdot}13} = -9{\cdot}538$ and $\alpha = 42^\circ$

Equations (128) and (130)
$$w_\mathrm{m}^\mathrm{W} = \frac{1{\cdot}220 \times 30{\cdot}50}{20\,000 \times 0{\cdot}61} + \frac{9 \times 0{\cdot}213 \times 3{\cdot}28^2}{20\,000 \times 0{\cdot}61^3}$$
$$= (3{\cdot}05 + 4{\cdot}54) \times 10^{-3}\ \mathrm{m}$$
$$= 7{\cdot}59\ \mathrm{mm}$$

Equations (129) and (130)
$$w_\mathrm{m}^\mathrm{T} = \mp 10^{-5} \times 20{\cdot}6 \times 30{\cdot}50 \mp \frac{9 \times 0{\cdot}443 \times 3{\cdot}28^2}{20\,000 \times 0{\cdot}61^3}$$
$$= \mp(6{\cdot}28 + 9{\cdot}44) \times 10^{-3}\ \mathrm{m}$$
$$= \mp 15{\cdot}72\ \mathrm{mm}$$

5.3.2. *Comparison with* in situ *measurements*[158]

The concrete strain at the dam's base was measured as $\varepsilon_x = \pm 115 \times 10^{-6}$. With Young's modulus of elasticity for concrete $E_\mathrm{c} = 20\ 000\ \mathrm{MN/m^2}$, the bending stress becomes $\sigma_x = \pm 2{\cdot}30\ \mathrm{MN/m^2}$ or 88% of the predicted value of $\pm 2{\cdot}62\ \mathrm{MN/m^2}$.

At midheight the downstream horizontal concrete strain has been measured as $\varepsilon_y = -275 \times 10^{-6}$ for the left abutment and $\varepsilon_y = -265 \times 10^{-6}$ for the right abutment. These values yield compressive stresses of $\sigma_{y\mathrm{A}}^\mathrm{W} = -5{\cdot}50$ and $-5{\cdot}30\ \mathrm{MN/m^2}$ respectively against a predicted value of $-5{\cdot}43\ \mathrm{MN/m^2}$. The measured downstream concrete strain for the crown of $\varepsilon_y = -75 \times 10^{-6}$ corresponds to a horizontal compressive stress of $\sigma_{y\mathrm{C}}^\mathrm{W} = -1{\cdot}50\ \mathrm{MN/m^2}$ compared to a predicted value of $-1{\cdot}90\ \mathrm{MN/m^2}$.

The radial crown displacement due to hydrostatic pressure was measured at midheight as $w_\mathrm{m}^\mathrm{W} = 8{\cdot}13$ mm compared to a predicted value of 7·59 mm.

If the shrinkage of the dam's four-month old concrete is estimated to be $\varepsilon_\mathrm{s} = 7 \times 10^{-5}$, the horizontal tensile stress amounts to

$$\sigma_y^\mathrm{S} = E_\mathrm{c}\varepsilon_\mathrm{s} = 20\,000 \times 7 \times 10^{-5} = 1{\cdot}40\ \mathrm{MN/m^2} \tag{137}$$

As this value exceeds the tensile strength of concrete, the observed vertical cracks in the middle of the dam are no surprise.

In total, a satisfactory agreement between *in situ* measurements and predictions by the approximate analysis can be noted.

5.4. Numerical example: Pacoima Dam (Fig. 85)

This 113·5 m high constant angle arch dam was designed by *Jorgensen* with help of the ring formula only[159] and built from 1926 to 1929 for flood control of Los Angeles, California. During the San Fernando Earthquake of 9 February 1971[82] the Pacoima Dam was exposed to a horizontal ground acceleration of $b_{\mathrm{Eh}} = 0{\cdot}7g$[81] which the dam sustained with little damage (opening of the construction joint between arch dam and left abutment block). Only the bedrock in the abutment regions required some strengthening. This result was due to a narrowing of the valley during the earthquake of about 24 mm.[160]

5.4.1. Approximate analysis

The values can be taken from Table 19. Because of the vertical curvature of the dam's upstream face the relieving vertical component of hydrostatic pressure N_x^{W} should not be overlooked.

5.4.2. Natural period of vibration

The radial crest displacement due to the 'fictional deadload acting in the horizontal direction' amounts approximately to

$$w = 2\left(\frac{\gamma_{\mathrm{c}}}{\gamma_{\mathrm{w}}}\right) w_{\mathrm{m}}^{\mathrm{W}} = 2\left(\frac{0{\cdot}025}{0{\cdot}01}\right) 5{\cdot}55 = 27{\cdot}75\ \mathrm{mm} \qquad (138)$$

and the natural period of vibration follows from Equation (36) as

$$T_1 = 1{\cdot}79\sqrt{0{\cdot}02775} = 0{\cdot}298\ \mathrm{s}$$

5.4.3. Improved equivalent static load

The four multipliers according to Equations (37) – (40) are

$$M_1 = \frac{0{\cdot}5}{0{\cdot}298} > 1{\cdot}0$$

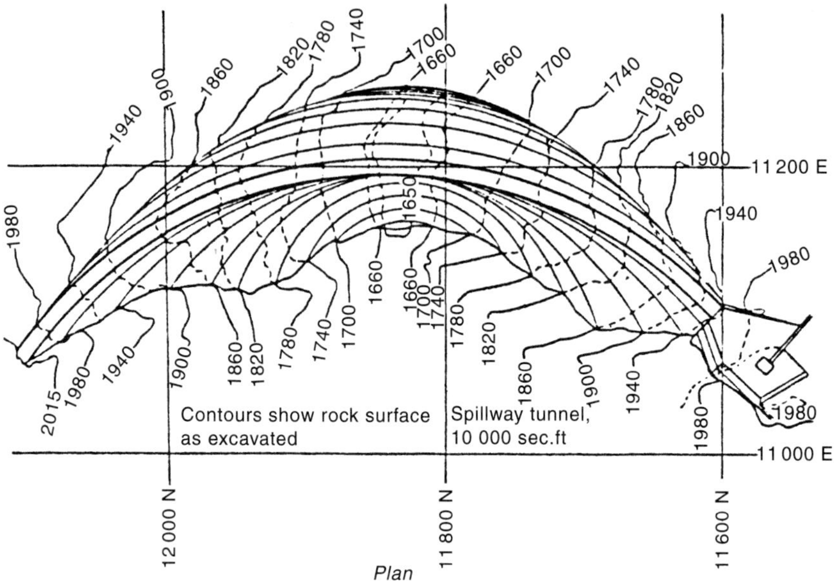

Plan
(a)

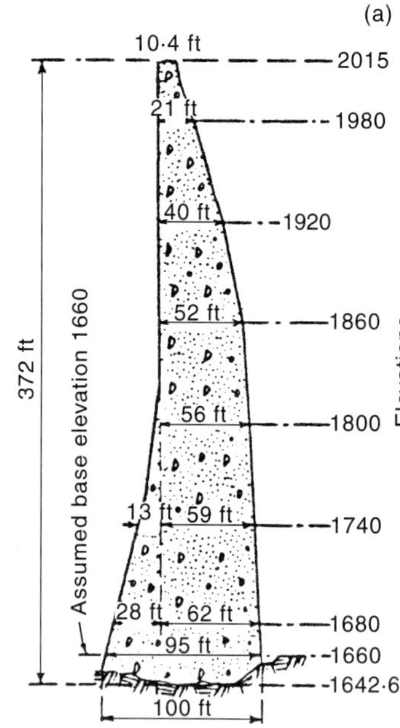

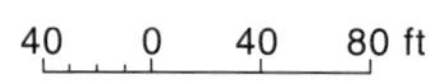

Elevation	Radius: ft		Thickness: ft	
	Extrados	Intrados	Crown	Abut
2015	335·80	325·40	10·4	10·4
1980	318·29	297·29	21·0	21·0
1920	281.55	232.85	40.0	43.4
1850	251·73	152·00	52·0	72·0
1800	234·47	111·30	59·7	78·0
1740	212·00	79·00	72·0	82·0
1680	171·34	64·00	90·0	
1660			95·0	
1642			100·0	

Maximum cross-section

Arch dimensions

(b)

Fig. 85. Dimensions of Pacoima constant angle arch dam in California: (a) *plan and* (b) *crown cantilever*[139]

$$M_2 = \left(\frac{2}{4{\cdot}477}\right)^2 = 0{\cdot}20$$

$$M_3 = 3/3 = 1{\cdot}0$$

$$M_4 = \frac{1}{2 \times 0{\cdot}05} = 10$$

These give (from Equation (42)) the ratio of the dam's response acceleration to the ground acceleration:

$$a_{\mathrm{Eh}}/b_{\mathrm{Eh}} = 1{\cdot}0 \times 0{\cdot}20 \times 1{\cdot}0 \times 10 = 2{\cdot}0$$

The seismic coefficient of the relevant transverse oscillation then follows from Equation (43) as

$$C_s = 2{\cdot}0 \times 0{\cdot}7 = 1{\cdot}4$$

This value is nearly ten times larger than the original design assumption of $C_s = 0{\cdot}15$.

5.4.4. Average horizontal ring stresses during the earthquake

At midheight the ring stresses listed in Table 20 were acting in the Pacoima Dam.

It follows from the above mentioned stress sums that the advantageous behaviour of the Pacoima Dam during the San Fernando Earthquake was solely due to the observed narrowing of the valley. From the compressive strength of concrete of $R_{\mathrm{c,stat}} = 34{\cdot}8\ \mathrm{MN/m^2}$ as measured in 1968, the static tensile strength of $R_{\mathrm{ct,stat}} = (34{\cdot}8/10)^{1/2} = 1{\cdot}87\ \mathrm{MN/m^2}$ can be deduced, giving a dynamic tensile strength of $R_{\mathrm{ct,dyn}} = 1{\cdot}25 \times 1{\cdot}87 = 2{\cdot}34\ \mathrm{MN/m^2}$. This value is greater than the highest tensile ring stress during the earthquake of $\sigma_y = 2{\cdot}05\ \mathrm{MN/m^2}$.

5.5. Numerical example: Hoover Dam (Fig. 18)

This 221 m high arch gravity dam in the Black Canyon of the Colorado River between Arizona and Nevada (Fig. 18) was an enormous step in dam development as the next smallest Owyhee Dam at $H = 127$ m was only half its height. Analysis of the Hoover Dam[53] by the then new trial load method[49] was checked for safety reasons

Table 19 (pages 148–153). Approximate analysis of 12 arch dams

Section	Name	Location	Construction	H: m	B_0: m	B_m: m	R_u: m	t_u: m	R_{mA}: m	t_{mA}: m	R_{mC}: m	t_{mC}: m
5.3	Stevenson Creek	USA	1926	18·3	42·7	23·52	29 · 66	2·29	30·50	0·61	30·50	0 · 61
5.4	Pacoima	USA	1926 – 29	113·5	195	109·9	36·6	30·5	57·15	22·88	57·15	16·47
5.5	Hoover	USA	1931 – 36	221·7	372	155	87·2	201·4	115·3	88·4	115·3	88·4
5.6	Lumiei	Italy	1944 – 47	136	138	81·65	20·0	15·6	30·20	11·29	36·55	10·26
5.7	Rossens	Switzerland	1945 – 48	83	320	244	106·0	28·0	134·4	17·6	136·2	14·07
5.8	Limberg	Austria	1947 – 51	120	350	194	66·6	39·1	126·3	19·40	126·3	19·40
5.9	Mauvoisin	Switzerland	1951 – 57	237	520	228	80·0	53·5	155·1	40·8	158·0	35·0
5.10	Gage	France	1953	37·6	142·9	73·4	22·2	2·57	64·3	1·40	64·3	1·40
5.11	Malpasset	France	1953 – 55	66·5	245	121	41·2	6·82	80·1	6·91	80·1	5·63
5.12	Zeuzier	Switzerland	1954 – 56	156[a]	256	194	60·6	25·6	83·1	15·6	83·1	15·6
5.13	Schiffenen	Switzerland	1961 – 63	47	417	372	197·5	13·0	200	8·00	200	8·00
5.14	Kölnbrein	Austria	1973 – 77	200	626	400	117	37·0	194	31·5	215	23·0

[a]Effective height only 122 m

Table 19. Continued

λ_x: m	λ_{yA}: m	λ_{yC}: m	Ec MN/m^2	E_r/E_c	N_x^G: MN/m	N_x^W: MN/m	M_x^W: MNm/m	Q_x^W: MN/m	P_y: MN/m^2	η	M_{yA}^W: MNm/m	N_{yA}^W: MN/m
6·26	3·28	3·28	20 000	1·0	−0·291	0	−2·36	0·95	0·040	0·443	−0·213	−1·220
25·39	23·32	23·32	20 000	1·0	−47·1	−2·48	−284	25·6	0·342	0·603	−93·0	−21·06
100·7	76·7	76·7	20 000	1·0	−525·3	−52·3	−12 270	172·5	0·330	0·298	−971	−52·7
13·42	14·03	14·72	20 000	1·0	−31·74	−4·95	−106·8	16·81	0·533	0·807	−52·4	−22·20
41·40	36·96	33·72	30 000	0·1	−30·9	−0·28	−35·65	2·58	0·384	0·925	−26·22	−54·99
38·77	37·63	37·63	25 000	0·75	−58·9	−2·85	−458	30·85	0·356	0·594	−189·1	−48·48
49·72	60·46	56·52	25 000	0·75	−82·0	−14·5	−1732	78·9	0·849	0·718	−1164	−149·0
5·74	7·21	7·21	18 000	0·75	−1·484	0	−3·94	1·496	0·1482	0·788	−2·89	−9·63
12·73	17·87	16·13	20 000	0·75	−8·55	−1·04	−27·81	4·9	0·2074	0·700	−24·84	−17·81
29·93	27·36	27·36	20 000	0·5	−47·5	−2·24	−206·2	16·02	0·479	0·785	−89·59	−43·51
38·51	30·40	30·40	40 000	0·1	−10·18	−0·18	−7·61	1·01	0·2136	0·909	−9·33	−43 · 57
50·00	59·41	53·44	20 000	1·0	−113·8	−6·1	−1875	87·5	0·5625	0·563	−992·7	−118·0

Table 19. Continued

y/λ_y	K	M^{W}_{yC}: MN m/m	N^{W}_{yC}: MN/m	T_x: degrees	T_y: degrees	M^{T}_{x}: MN m/m	M^{T}_{yA}: MN m/m	N^{T}_{yA}: MN/m	M^{T}_{yC}: MN m/m	N^{T}_{yC}: MN/m	σ^{G}_{x}: MN/m^2 US[a]	DS[b]
3·59	0·014	0·0043	−1·220	13·2	20·6	3 · 99	0·443	−2·51	−0·0089	−2·51	−0·25	0
2·36	0·134	25·1	−21·06	1·9	3·3	102·1	51·72	−15·10	−13·96	−10·87	−2·57	−0·51
1·01	0·208	404	−52·7	0·30	0·68	703	307	−12·02	−127·6	−12·02	−3·96	−1·26
2·77	0·070	7·34	−22·20	3·4	4·9	47·74	36·06	−11·06	−4·77	−10·08	−4·07	0
3·30	−0·006	−0·315	−54·99	2·0	3·9	13·58	10·20	−20·05	0·122	−16·04	−2·21	0
2·58	0·120	47·4	−48·48	1·5	2·8	165·4	57·1	−13·58	−13·70	−13·58	−3·01	0
2·02	0·165	384	−149·0	1·1	1·7	205	135	−17·34	−50·6	−14·88	−3·06	0
5·09	0	0	−9·63	12 · 5	16·3	3·22	1·245	−4·11	0	−4·11	−1·15	0
3·39	0·015	0·847	−17·81	6·7	7·7	13·49	15·92	−10·64	−0·478	−8·67	−1·63	−0·87
3·55	0·015	2·69	−43·51	2·2	3·4	41·62	23·89	−10·61	−0·717	−10·61	−3·77	0
6·12	0	0	−43·57	5·2	6·0	7·42	4·44	−19·20	0	−19·20	−1·57	0
3·20	0·040	79·4	−127·4	1·6	1·8	126·5	103·1	−11·34	−8·25	−8·28	−7·26	1·11

[a] US = upstream face
[b] DS = downstream face

Table 19. Continued

e^{W}: m	ΔM_x^{W}: MN m/m	Portion cf due to		σ_x^{W}: MN/m^2		σ_x^{G+W}: MN/m^2		σ_{x0}^{G+W}: MN/m^2	τ_x^{W}: MN/m^2	σ_1: MN/m^2	σ_2: MN/m^2	α degrees
		N_x^{W}	$M_x^{W} + \Delta M_x^{W}$	US	DS	US	DS					
0	0	0	±2·62	2·62	−2·62	2·37	−2·62	−0·13	0·62	0·56	−0·70	42
10·22	25·3	−0·08	±1·67	1·59	−1·75	−0·98	−2·26	−1·62	1·26	0·69	−2·31	29
86·52	4525	−0·26	±1·15	0·89	−1·41	−3·07	−2·67	−2·87	1·28	0·49	−3·36	21
2·57	12·72	−0·32	±2·32	2·00	−2·64	−2·07	−2·64	−2·36	1·62	0·82	−3·18	27
10·74	3·01	−0·01	±0·25	0·24	−0·26	−1·97	−0·26	−1·12	0·14	0·52	−1·14	7
15·17	43·2	−0·07	±1·63	1·56	−1·70	−1·45	−1·70	−1·58	1·18	0·63	−2·21	28
19·85	287·8	−0·27	±3·03	2·76	−3·30	−0·30	−3·30	−1·80	2·21	1·48	−3·28	34
0	0	0	±3·58	3·58	−3·58	2·43	−3·58	−0·58	0·87	0·63	−1·21	36
2·21	2·30	−0·15	±3·29	3·14	−3·44	1·51	−4·31	−1·40	1·09	0·60	−2·00	29
10·71	23·99	−0·09	±1·89	1·80	−1·98	−1·97	−1·98	−1·98	0·94	0·38	−2·36	22
5·90	1·06	−0·01	±0·23	0·22	−0·24	−1·35	−0·24	−0·80	0·12	0·02	−0·82	8
17·05	104·0	−0·16	±7·76	7·60	−7·92	0·34	−6·81	−3·24	3·55	2·28	−5·52	33

Table 19. Continued

Portion of due to		σ^{W}_{yA}: MN/m²		Portion of due to		σ^{W}_{yC}: MN/m²		Portion of due to		
N^{W}_{yA}	M^{W}_{yA}	US	DS	N^{W}_{yC}	M^{W}_{yC}	US	DS	N^{W}_{yC}	M^{W}_{yA}	w^{W}_{m}: mm
−2·00	±3·43	1·43	−5·43	−2·00	∓0·10	−2·10	−1·90	3·05	4·54	7·59
−0·92	±1·07	0·15	−1·99	−1·28	∓0·55	−1·83	−0·73	3·65	1·90	5·55
−0·60	±0·75	0·15	−1·35	−0·14	∓0·24	−0·38	0·10	3·43	3·72	7·15
−1·97	±2·47	0·50	−4·44	−2·16	∓0·42	−2·58	−1·74	3·95	3·22	7·17
−3·12	±0·51	−2·61	−3·63	−3·91	∓0·01	−3·92	−3·90	17·74	1·98	19·72
−2·50	±3·02	0·52	−5·52	−2·50	∓0·75	−3·25	−1·75	12·63	13·20	25·83
−3·65	±4·20	0·55	−7·85	−4·26	∓1·88	−6·14	−2·38	26·91	22·55	49·46
−6·88	±8·85	1·97	−15·73	−6·88	0	−6·88	−6·88	24·57	27·38	51·95
−2·58	±3·12	0·54	−5·70	−3·16	∓0·16	−3·32	−3·00	12·66	10·81	23·47
−2·79	±2·24	−0·55	−5·03	−2·79	∓0·07	−2·86	−2·72	11·59	7·95	19·54
−5·45	±0·87	−4·58	−6·32	−5·45	0	−5·45	−5·45	27·23	3·79	31·02
−3·75	±6·00	2·25	−9·75	−5·54	∓0·90	−6·44	−4·64	59·55	50·45	110·00

Table 19. Continued

σ_x^T: MN/m^2	Portion of due to		σ_{yA}^T: MN/m^2		Portion of due to		σ_{yC}^T: MN/m^2		Portion of due to		
	N_{yA}^T	M_{yA}^T	US	DS	N_{yC}^T	M_{yC}^T	US	DS	N_{yC}^T	M_{yA}^T	w_m^T: mm
∓4·57	−4·12	∓7·14	−11·26	3·02	−4·12	±0·20	−3·92	−4·32	∓6·28	∓9·44	∓15·72
∓0·66	−0·66	∓0·59	−1·25	−0·07	−0·66	±0·31	−0·35	−0·97	∓1·88	∓1·06	∓2·94
∓0·10	−0·14	∓0·24	−0·38	0·10	−0·14	±0·10	−0·04	−0·24	∓0·70	∓1·17	∓1·87
∓1·18	−0·98	∓1·70	−2·68	0·72	−0·98	±0·24	−0·74	−1·22	∓1·79	∓2·22	∓4·01
∓0·10	−1·14	∓0·20	−1·34	−0·94	−1·14	0	−1·14	−1·14	∓5·18	∓0·77	∓5·95
∓0·65	−0·70	∓0·91	−1·61	0·21	−0·70	±0·03	−0·67	−0·73	∓3·54	∓3·98	∓7·52
∓0·36	−0·43	∓0·55	−0·98	0·12	−0·43	±0·18	−0·25	−0·61	∓2·69	∓2·97	∓5·66
∓2·93	−2·93	∓3·81	−6·74	0·88	−2·93	0	−2·93	−2·93	∓10·56	∓11·80	∓22·36
∓1·74	−1·54	∓2·00	−3·54	0·46	−1·54	±0·09	−1·45	−1·63	∓6·16	∓6·93	∓13·09
∓0·38	−0·68	∓0·59	−1·27	−0·09	−0·68	±0·02	−0·66	−0·70	∓2·83	∓2·12	∓4·95
∓0·26	−2·40	∓0·42	−2·82	−1·98	−2·40	0	−2·40	−2·40	∓12·00	∓1·80	∓13·80
∓0·55	−0·36	∓0·62	−0·98	0·26	−0·36	±0·05	−0·31	−0·41	∓3·87	∓5·24	∓9·11

Table 20. Ring stresses in Pacoima Dam

No.	Loading	σ_y: MN/m^2
1	Deadload	0
2	Hydrostatic pressure (water level 45 m below crest)	−0·44
3	Earthquake (horizontal component of deadload)	±6·44
4	Earthquake (horizontal component of hydrostatic pressure for water level 45 m below crest)	±0·86
5	Narrowing of valley by 24 mm	−4·80
6	1+3+5 (reservoir empty)	1·61 −11·24
7	1+2+3+4+5 (reservoir partly filled with water level 45 m below crest)	2·05 −12·54
8	1+2+3+4 (as before but without narrowing of valley)	6·85 −7·74

with a model test[54] at the University of Boulder, Colorado. The model to the scale of 1 in 240 was fabricated from a mixture of plaster, diatomaceous earth and water, and loaded with mercury.

5.5.1. *Approximate analysis*

The corresponding values can again be taken from Table 19. The deadload stresses were computed for the Hoover Dam acting as a rigid body (linear stress distribution in horizontal sections). As the 201 m thick dam was constructed of a large number of individual blocks (square concrete columns) the distribution of deadload stresses is actually proportional to the block heights.

5.5.2. *Comparison with model test*[54]

The following values due to hydrostatic pressure were measured at midheight.

(a) Radial displacement $w_{\mathrm{m}}^{\mathrm{W}} = 8{\cdot}81$ mm as compared to the prediction by approximate analysis of 7·15 mm.
(b) Downstream strains in horizontal direction at the abutment $\varepsilon_{y\mathrm{A}}^{\mathrm{W}} = -17 \times 10^{-6}$ (Arizona side) and $\varepsilon_{y\mathrm{A}}^{\mathrm{W}} = -39 \times 10^{-6}$ (Nevada side) with an average of $\varepsilon_{y\mathrm{A}}^{\mathrm{W}} = -28 \times 10^{-6}$ as well as $\varepsilon_x^{\mathrm{W}} = -39 \times 10^{-6}$ at the base.

For the presumed Young's modulus of $E_c = 20\,000$ MN/m^2, the downstream horizontal compressive stress at midheight amounts to $\sigma_{yA}^{W} = -0{\cdot}56$ MN/m^2 (prediction $-1{\cdot}35$ MN/m^2) at the abutment and the vertical compressive stress at the dam base to $\sigma_x^{W} = -0{\cdot}78$ MN/m^2 (prediction $-1{\cdot}41$ MN/m^2).

Bearing in mind how little the massive concrete plug of the Hoover Dam (its concrete volume slightly exceeds the stone volume of Cheop's pyramid) resembles the static system of the approximate analysis (circular cylindrical tank of constant thickness) the close agreement between radial displacements according to model tests and those determined from the approximate analysis is rather astonishing.

5.5.3. Conversion of model results to the actual Hoover Dam

In the conversion of the model results to the full-size dam the following known model laws[161] must be observed.

(a) Loads and stresses

$$\frac{\sigma}{\sigma_M} = \frac{n}{\gamma_M} = \frac{240}{13{\cdot}6} = 17{\cdot}65 \tag{139}$$

(b) Strains

$$\frac{\varepsilon}{\varepsilon_M} = \frac{nE_M}{\gamma_M E} = \frac{240}{13{\cdot}6} \times \frac{630}{20\,000} = 0{\cdot}556(\times 39 \times 10^{-6} = 22 \times 10^{-6}) \tag{140}$$

(c) Displacements

$$\frac{w}{w_M} = \frac{n^2 E_M}{\gamma_M E} = \frac{240^2}{13{\cdot}6} \times \frac{630}{20\,000} = 133{\cdot}4\left(\times \frac{25{\cdot}4 \times 26}{10\,000} = 8{\cdot}81 \text{ mm}\right) \tag{141}$$

5.6. Numerical example: Lumiei Dam (Fig. 86)

This 132 m high arch dam built in the Carnic Alps in Italy from 1944 to 1947 was at that time Europe's highest dam and this with a perimetral joint (Fig. 86).[162]

5.6.1. Approximate analysis

For the corresponding values see Table 19.

5.6.2. Comparison with model test[162]

On a second model (scale 1 in 60) the radial displacement due to hydrostatic pressure at midheight was measured as $E_0 w_m^W =$ 133 MN/m. Then, Young's modulus of $E_c = 20\ 000$ MN/m^2 yields

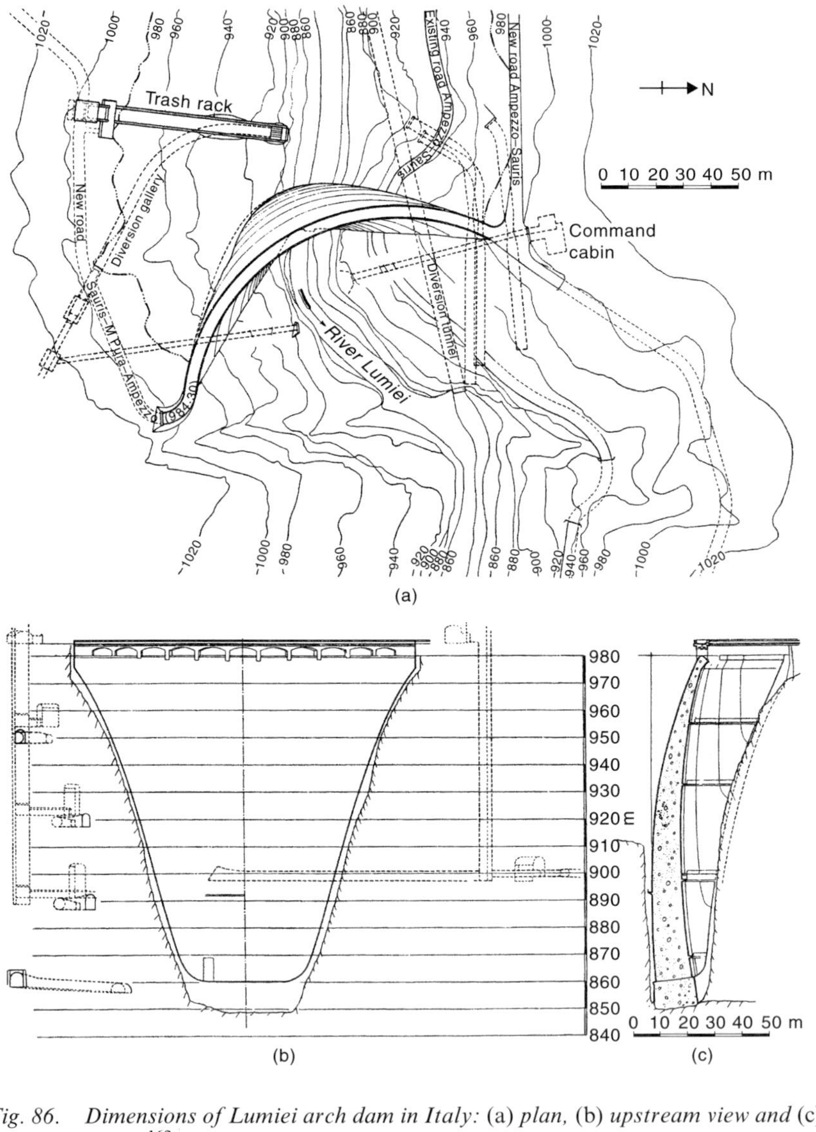

Fig. 86. Dimensions of Lumiei arch dam in Italy: (a) *plan,* (b) *upstream view and* (c) *crown cantilever*[162]

$w_m^W = 6{\cdot}65$ mm whilst the approximate analysis gives 7·17 mm. The measured horizontal stresses (Table 21) due to the same loading equally at midheight show considerable differences to the predictions (in brackets in Table 21) by the approximate analysis.

The dying away of the horizontal bending moments from the abutment

$$M_{yA}^W = -(3{\cdot}27 - 0{\cdot}53)\left(\frac{11{\cdot}29^2}{6}\right)$$
$$= -58{\cdot}2 \text{ MN/m (prediction} - 52{\cdot}4 \text{ MN m/m)}$$

to the crown

$$M_{yC}^W = (1{\cdot}48 - 0{\cdot}82)\left(\frac{10{\cdot}26^2}{6}\right)$$
$$= 11{\cdot}23 \text{ MN/m (prediction } 7{\cdot}34 \text{ MN m/m)}$$

is clearly recognizable. The measured horizontal direct forces at the abutment

$$N_{yA}^W = -(3{\cdot}27 + 0{\cdot}53) \times 11{\cdot}29 = -42{\cdot}90 \text{ MN/m}$$

and at the crown

$$N_{yC}^W = -(1{\cdot}48 + 0{\cdot}82) \times 10{\cdot}26 = -23{\cdot}60 \text{ MN/m}$$

coincide with the prediction ($N_y^W = -22{\cdot}20$ MN/m) only for the crown.

5.6.3. *Rupture test of the model*[162]

Loading of the model in the rupture test with small hydraulic jacks simulated a fictional fluid with a density of $\gamma_u = 8 \times 0{\cdot}089 = 0{\cdot}712$ MN/m^3. The model scale of 1 in 60, the compressive strength

Table 21. Measured (and predicted) horizontal stresses in Lumiei Dam

	σ_y^W: MN/m^2			
Face of dam	abutment		crown	
Upstream	−0·53	(0·50)	−1·48	(−2·58)
Downstream	−3·27	(−4·44)	−0·82	(−1·74)

of the plaster – diatomaceous earth mortar of $R_M = 3{\cdot}5\ \mathrm{MN/m^2}$ and that of the dam's concrete of $R_c = 35\ \mathrm{MN/m^2}$ yields the fictional density of the model fluid which corresponds to natural water as $\gamma_M = 0{\cdot}01 \times 60\ (3{\cdot}5/35) = 0{\cdot}060\ \mathrm{MN/m^3}$. Thus, the safety of rupture amounted to

$$s = \frac{\gamma_u}{\gamma_M} = \frac{0{\cdot}712}{0{\cdot}060} = 11{\cdot}9 \tag{142}$$

This value corresponds approximately to the ratio of the dam's concrete strength of $R_c = 35\ \mathrm{MN/m^2}$ to the horizontal direct stress at midheight[163] by the ring formula (Equation (108)) of

$$\begin{aligned}\sigma_y^W &= -0{\cdot}01\left(\frac{136}{2}\right)\left(36{\cdot}55 + \frac{10{\cdot}26}{2}\right)/10{\cdot}26 \\ &= -2{\cdot}76\ \mathrm{MN/m^2}\end{aligned}$$

Thus

$$s = 35/2{\cdot}76 = 12{\cdot}7$$

5.7. Numerical example: Rossens Dam (Fig. 87)

This 83 m high arch dam with a crest length of 320 m ($B/H = 3{\cdot}86$) on the Sarine River in the Swiss Canton of Fribourg was built from 1945 to 1948.[164] The peculiarity of this dam lies in the bedrock. The modulus of elasticity of the marine sandstone (molasse) is one order of magnitude smaller than that of the dam's concrete.

5.7.1. *Approximate analysis*

For the corresponding values, again see Table 19.

5.7.2. *Comparison with* in situ *measurements*

On 22 May 1962 the radial displacement at midheight was measured for a full reservoir to be $27 - 9 = 18$ mm (Ref. 137, p. 175, Fig. 10). At this time of the year a slight increase of concrete temperature (about 10% of the annual amplitude) has already taken place. Then, the radial displacement according to the approximate analysis amounts to

$$w_{\mathrm{m}}^{\mathrm{W+T/10}} = 19{\cdot}72 - \frac{5{\cdot}95}{10} = 19{\cdot}12 \text{ mm}$$

The agreement between *in situ* measurement and prediction is remarkable.

5.8. Numerical example: Limberg Dam (Fig. 88)

This 120 m high asymmetrical arch dam on the Kapruner Ache Torrent in the High Tauern Region of the Austrian Province of Salzburg was built in 1947 – 1951. It creates a reservoir of 88×10^6 m^3. The bedrock consists of calcareous mica schists. The trough-like cross-section of the valley flattens out on the left flank (Fig. 88). The Limberg Dam was the first arch dam analysed for six arches and six cantilevers by the algebraic load distribution method[55] instead of the cumbersome trial load method.[49] It requires solution of a system of algebraic equations for the unknown displacements at the intersections of arches and cantilevers.

5.8.1. Approximate analysis

The corresponding values can again be taken from Table 19. As the asymmetry of the valley's cross-section was disregarded, the predicted fixing moments at the abutments are of the same size for both valley flanks.

5.8.2. Comparison with the design analysis

Tremmel[165] communicated the stresses due to hydrostatic pressure computed for radial adjustment only. For the arch at midheight a linear interpolation yields the values listed in Table 22.

The approximate analysis yields somewhat different values as shown in Table 23.

The differences in values listed in Tables 22 and 23 originate from different distributions of horizontal bending moments at midheight (Table 24).

The reduction of horizontal bending moments from the abutment to the crown is less pronounced for the design analysis (based on a grid of six arches and six cantilevers[55]) than for the approximate analysis (based on the theory of shells[64]).

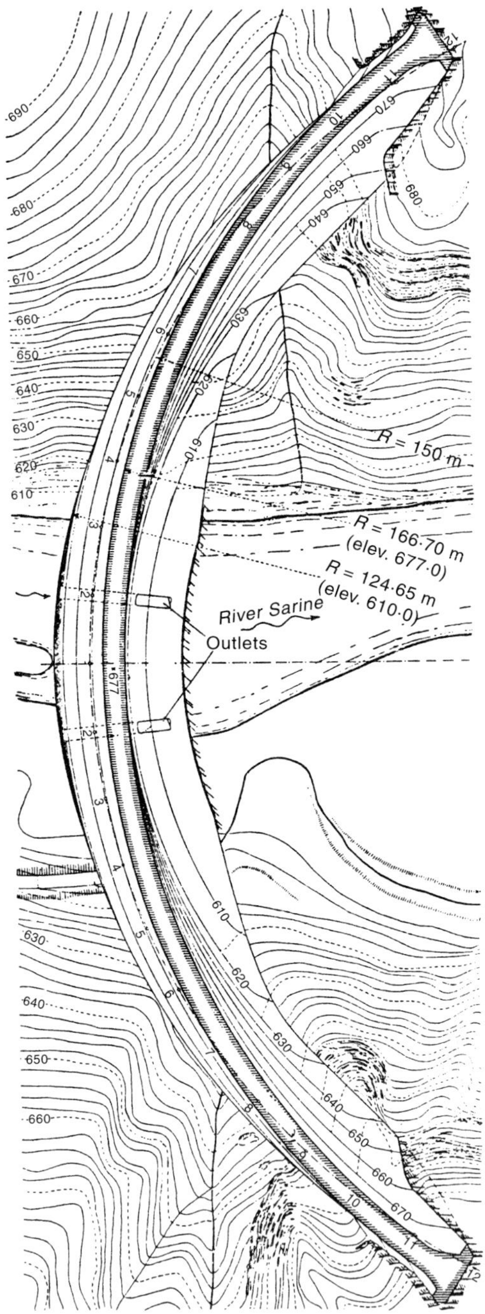
R = 150 m
R = 166·70 m
(elev. 677·0)
R = 124·65 m
(elev. 610·0)
River Sarine
Outlets
677

(a)

5.8.3. Comparison with *in situ* measurements

Allowing for the phase difference between air and concrete temperatures of about one month for each 10 m of wall thickness, long-term observations of the Limberg Dam[167] have shown that the largest radial displacement due to hydrostatic pressure coincides with an increase in concrete temperature in autumn. The largest crest displacement of 30 mm corresponds to a midheight value of about 18 mm, whilst the approximate analysis yields $w_{\mathrm{m}}^{\mathrm{W+T}} = 25{\cdot}83 - 7{\cdot}52 = 18{\cdot}31$ mm.

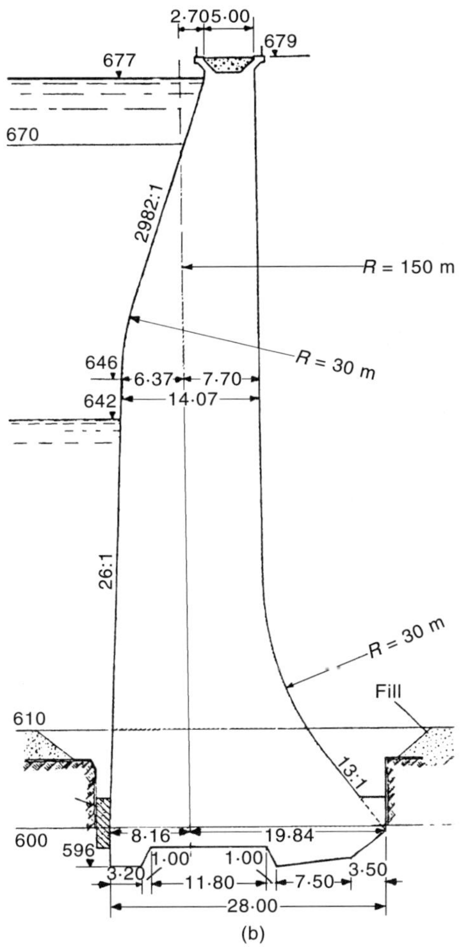

Fig. 87 (facing and above). Dimensions of Rossens arch dam in Switzerland: (a) *plan and* (b) *vertical section*[164]

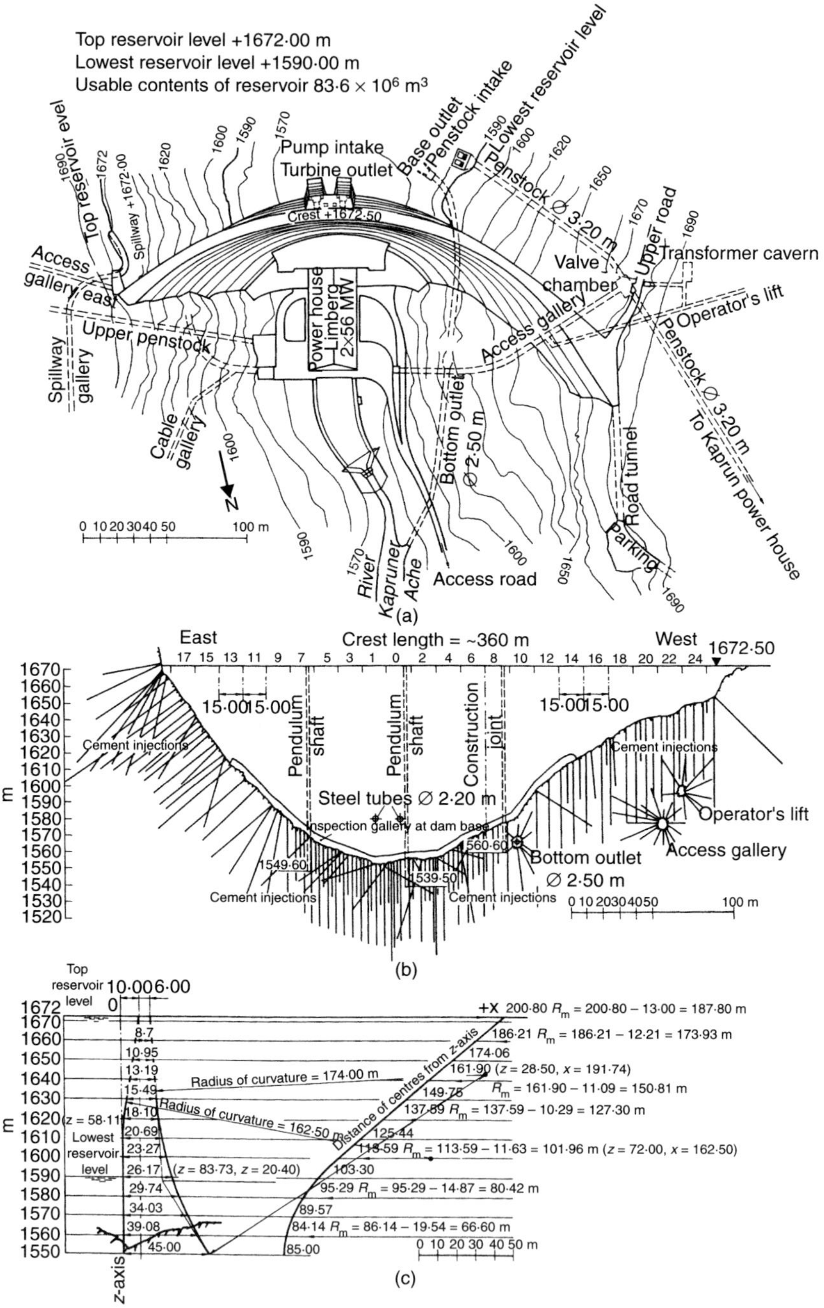

Fig. 88. Dimensions of Limberg constant angle arch dam in Austria: (a) *plan,* (b) *downstream view and* (c) *crown cantilever*[166]

Table 22. Computed stresses due to hydrostatic pressure on Limberg Dam

	σ_y^W: MN/m^2	
Cross-section	Upstream	Downstream
Abutment east	0·37	−3·97
Crown	−3·80	−0·28
Abutment west	0·01	−3·25

Table 23. Approximate stresses due to hydrostatic pressure on Limberg Dam

	σ_y^W: MN/m^2	
Cross-section	Upstream	Downstream
Abutment	0 · 52	−5 · 52
Crown	−3 · 25	−1 · 75

Table 24. Distribution of horizontal bending moments (in MN m/m) at midheight for Limberg Dam

	Abutment		
Analysis	East	West	Crown
Design	−147	−213	119
Approximate	−189	−189	47

5.9. Numerical example: Mauvoisin Dam (Fig. 89)

This 237 m high constant angle arch dam with a crest length of 520 m was built in the Bagnes Valley in the Swiss Canton of Valais from 1951 to 1957.[168] Its wall thickness is 14·0 m at the crest and 53·5 m at the base. The bedrock of homogeneous calcareous mica schists was grouted to a depth of 215 m below the dam's base. 247 000 m^2 of grout curtain required 8050 tonnes of cement. The mass concrete contains 175 kg of cement per cubic metre and the facing concrete (4·5 m thick on the upstream face and 3·5 m thick on the downstream face) contains 250 kg/m^3. The concrete was artificially cooled with water circulating in pipes.

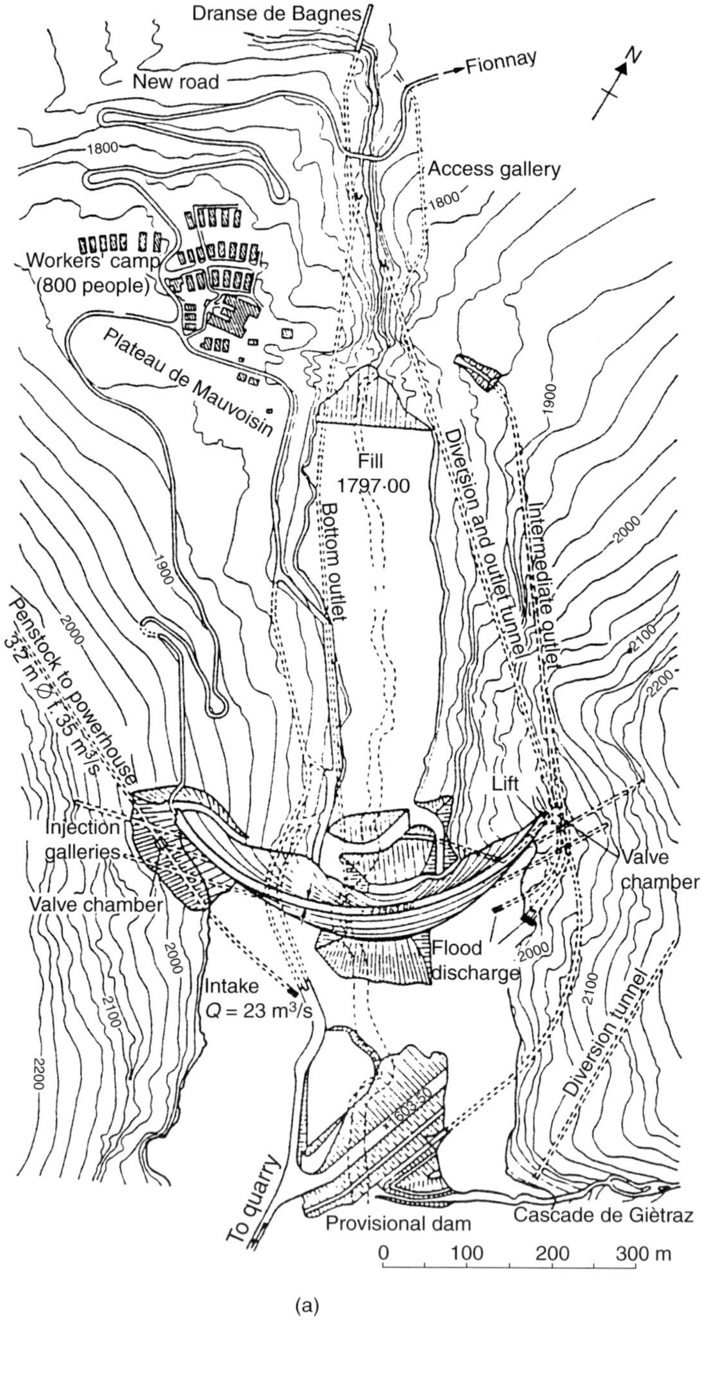
Dranse de Bagnes
New road
Fionnay
N
1800
Access gallery
1800
Workers' camp
(800 people)
Plateau de Mauvoisin
1900
Fill
1797·00
Diversion and outlet tunnel
Intermediate outlet
Bottom outlet
2000
1900
2100
2200
Penstock to powerhouse
3.2 m Ø f. 35 m³/s
2000
Lift
Injection
galleries
Valve
chamber
Valve chamber
2000
Flood
discharge
2000
2100
Intake
Q = 23 m³/s
2100
Diversion tunnel
2200
To quarry
Provisional dam
Cascade de Giètraz
0
100
200
300 m

(a)

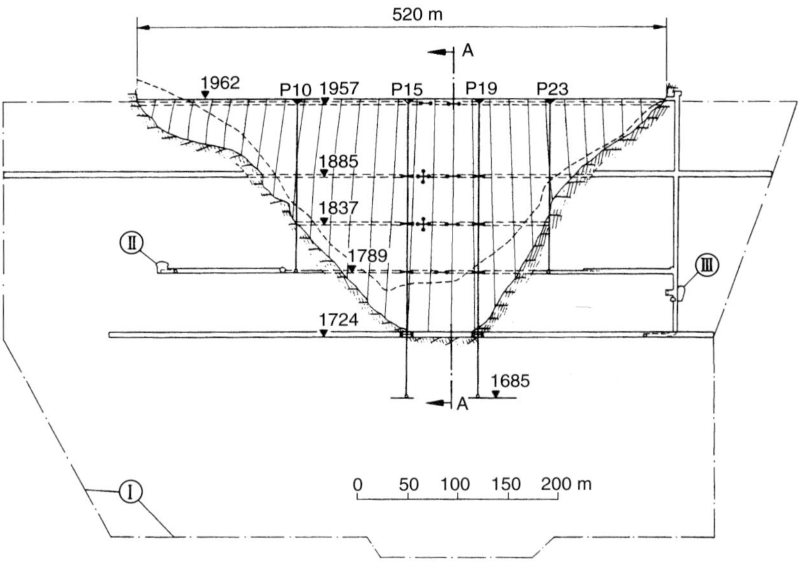

(b)

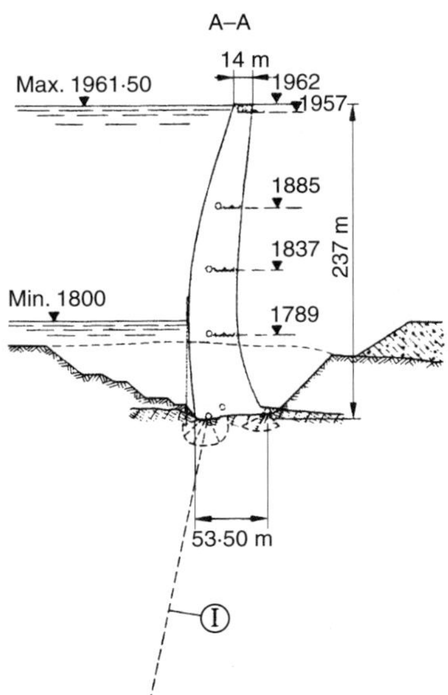

(c)

Fig. 89 (facing and above). Mauvoisin constant angle arch dam in the Hérémence Valley, Switzerland, before its heightening: (a) *plan,* (b) *upstream view (I = grout curtain, II = valve chamber of pressure tunnel intake, III = valve chamber of bottom outlet) and* (c) *vertical section (block no. 15)*[171]

5.9.1 Approximate analysis

For the corresponding values again see Table 19.

5.9.2. Observations and measurements

Observation of the dam's behaviour began during construction.[169] Four plumb-lines, 34 clinometers, 38 strainmeters, 14 upliftmeters and 22 seepage-meters were installed. Many electric thermometers are no longer in action, but some of them could be activated if necessary. A comparison of radial displacements over a period of four decades shows that the largest crest displacement was of about the same size up to 1962, then it increased slowly up to 1967 (e.g. displacement of block no. 15 increased from 75 to 95 mm) and since then it remained practically constant for 23 years until the dam's heightening in 1990 – 1991. The seepage diminished from 320 l/s in autumn 1964 to 110 l/s in autumn 1982. The uplift pressure has been practically constant since 1958.[78,101,131,168,169]

Increases in radial displacements from 1960 to 1977 (see Ref. 170, p. 915, Fig. 4) of 2 mm at the base and 19 mm at the crest are due firstly to plastic deformations of the bedrock and secondly to concrete shrinkage. As the artificial cooling was stopped in spring 1958 the increase in concrete temperature due to a build-up of the hydration heat of cement up to 1962 partially compensated the concrete shrinkage during this period.

The measured vertical displacements of the crest permit an estimate of concrete shrinkage (Ref. 78, p. 636, Fig. 5). The heave of 11 mm measured in October 1958 diminished to 8 mm in September 1961. Assuming a parabolic distribution of shrinkage over the dam's height (the concrete at the base does not shrink at all) the heave difference of 3 mm in three years (at equal reservoir levels) yields a specific shrinkage of

$$\Delta\varepsilon_s = \frac{3u}{H} = \frac{-3 \times 3}{237\,000} = -3{\cdot}8 \times 10^{-5} \tag{143}$$

As the heave values have remained practically constant since 1967 shrinkage of the dam must have terminated in 1967. It can thus be estimated that total shrinkage occurring during the ten years after completion of the dam will amount to about twice the value for the first three years ($\varepsilon_s = -7{\cdot}6 \times 10^{-5}$).

The behaviour of block joints is also very interesting. In the crest region they open if the reservoir level is lowered and close if it rises. The largest concrete strain was measured between the highest blocks nos 17 and 18 at El. 1957 m (or 5 m below crest) as $\varepsilon_c = 0{\cdot}25/600 = 417 \times 10^{-6}$. With a Young's modulus $E_c = 25\ 000$ MN/m^2, the corresponding tensile stress of $\sigma_y = 10{\cdot}4$ MN/m^2 is far greater than the tensile strength of concrete. The block joints thus had to open.

5.9.3. *Comparison with* in situ *measurements*

5.9.3.1. *Radial displacements*

The radial displacements of block no. 15 were measured for the same reservoir level at El. 1817 m in May 1960 and 1972 as well as for the same reservoir level at El. 1959 m in October of the same years (Ref. 170, p. 915, Fig. 4). The differences in measured radial displacements (Table 25) can be attributed to small differences in concrete temperature in the two years of measurement, e.g. for the crest

$$\Delta T = \frac{\Delta w}{\omega R_y} = \frac{-4}{10^{-5} \times 307\,000} = -1{\cdot}3\,^{\circ}\mathrm{C} \tag{144}$$

From the measured radial displacements (including the plastic deformation of bedrock) of block no. 15 for a full reservoir in October 1982 (Fig. 90, after Ref 101, p. 43, Figs 4.1.1 – 4) the value at midheight (El. 1843 m) of $w_m^{W+T} = 51$ mm can be obtained. After substraction of the portion due to the plastic deformation of bedrock of 8 mm (Ref 170, p. 915, Fig. 4) the remaining elastic portion amounts to $w_{m,el}^{W+T} = 51 - 8 = 43$ mm. The approximate analysis with $E_c = 25\ 000$ MN/m^2, with consideration of the still warm concrete in autumn, yields the value of

$$w_m^{W+T} = 49{\cdot}46 - 5{\cdot}66 = 43{\cdot}80 \text{ mm}$$

Table 25. Radial displacements of Mauvoisin Dam measured in 1960 and 1972

	1960			1972			
Elevation: m	May	October	w_{60}: mm	May	October	w_{72}: mm	$\Delta w = w_{72} - w_{60}$: mm
1958.6	16	75	59	31	94	63	4
1922.6	13	72	59	25	85	60	1
1886.6	12	65	53	22	76	54	1
1838.6	9	49	40	18	58	40	0
1790.6	7	27	20	11	34	23	3
1726.4	1	3	2	2	5	3	1

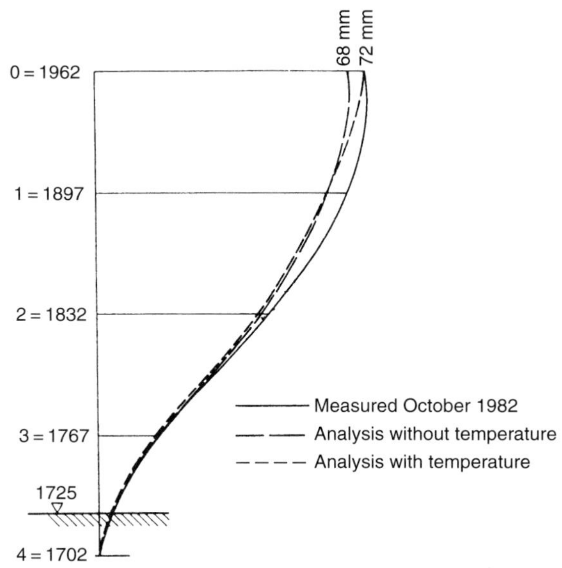

Fig. 90. Radial displacements of block no. 15 for a full reservoir in October 1982[101] *compared with approximate analyses*[171]

5.9.3.2. Vertical bending

The curvature of the line of radial displacements (Fig. 90)

$$\frac{1}{R_x} = \frac{d^2w}{dx^2} = w'' \tag{145}$$

allows calculation of bending moments with the help of the well known formula

$$M = -EIw'' \tag{146}$$

Assuming $E_c = 25\,000$ MN/m^2, the fixing moment at the base of block no. 15 is

$$M_x = -25\,000\left(\frac{53 \cdot 5^2}{12}\right)7 \cdot 50 \times 10^{-6} = -2393 \text{ MN m/m}$$

whilst the approximate analysis for $E_r = 0 \cdot 75E_c$ yields $M_x = -1732$ MN m/m and for $E_r = E_c$ the value of M_x is -2305 MN m/m.

A check is possible with help of Equation (60) for the dam's base rotation which, for $E_r = 0{\cdot}75E_c = 18\ 750\ \mathrm{MN/m^2}$, is

$$\alpha = \frac{18 \times 2393}{\pi 18\,750 \times 53{\cdot}5^2} = 255 \times 10^{-6}$$

and for $E_r = E_c = 25\ 000\ \mathrm{MN/m^2}$ is $\alpha = 340 \times 10^{-6}$. From the line of radial displacements (Figure 90) the value of $\alpha = 243 \times 10^{-6}$ can be taken.

5.9.3.3. *Vertical displacements*

The different portions of vertical displacement[171] are as follows.

(a) Transverse expansion due to the horizontal ring force resulting from hydrostatic pressure and due to Poisson's ratio

$\mu = \frac{1}{6} = 0{\cdot}167$

$$u_\mu = \frac{\mu}{H}\int_0^H w\ \mathrm{d}x = \mu w_\mathrm{m}^\mathrm{W} = \frac{43{\cdot}8}{6} = 7{\cdot}3\ \mathrm{mm} \qquad (147)$$

(b) Shrinkage of concrete

$$u_\mathrm{S} = \frac{\varepsilon_\mathrm{S} H}{3} = \frac{7{\cdot}6}{3} \times 10^{-5} \times 237\,000 = 6{\cdot}0\ \mathrm{mm} \qquad (148)$$

(c) Variation of temperature according to Equation (1) of $T = 62/(14{\cdot}0 + 2{\cdot}4) = 3{\cdot}8\ °\mathrm{C}$ at the crest

$$u_\mathrm{T} = \frac{\omega T H}{3} = 10^{-5}\left(\frac{3{\cdot}8}{3}\right)237\,000 = 3{\cdot}0\ \mathrm{mm} \qquad (149)$$

If the shrinkage of concrete is at first left aside, the approximate analysis with the help of Equations (147) and (149) yields $u = 7{\cdot}3 + 3{\cdot}0 = 10{\cdot}3$ mm. For a full reservoir and still warm concrete the value of $u = 11$ mm was measured in October 1958 (Ref. 78, p. 636, Fig. 5). Until September 1961 this heave diminished by shrinkage of concrete to $u = 10{\cdot}3 - (6{\cdot}0/2) = 7{\cdot}3$ mm (*in situ* measurement 8 mm). Up to the presumed termination of shrinkage in 1967, the crest heave for a full reservoir diminished to $u = 10{\cdot}3 - 6{\cdot}0 = 4{\cdot}3$ mm.

5.9.3.4. Deformations of the valley

Three characteristic values must be distinguished which depend on the following four parameters:

(a) height of the dam equalling the depth of the reservoir;
(b) developed length of dam base a;
(c) length of reservoir L_s;
(d) modulus of elasticity of bedrock E_r.

The *settlement of the valley floor* due to the reservoir load of $1{\cdot}8 \times 10^6$ MN for $E_r = 18\ 750\ \text{MN/m}^2$ is (Equation (64))

$$\max\ u_r = 0{\cdot}39\left(\frac{0{\cdot}01 \times 237 \times 750}{18\,750}\right) = 0{\cdot}0370\ \text{m}$$

The *widening of the valley* reaches its maximum in the lower quarter-point of the dam's height. From Equation (65), and depending on the inclination of valley flanks (left side $\beta = 40°$ and right side $\beta = 50°$, average value $\beta = 45°$) it is

$$\max\ v_r = 0{\cdot}50\left(\frac{0{\cdot}01 \times 237 \times 750}{18\,750}\right)0{\cdot}707^2 = 0{\cdot}0237\ \text{m}$$

Rotation of the valley floor below the dam equals the inclination of the settlement bowl

$$u'_r = \frac{\max\ u_r}{2H} = \frac{0{\cdot}0370}{2 \times 237} = 78 \times 10^{-6} \qquad (150)$$

This value is

$$\frac{78}{243 + 78} \times 100 = 24{\cdot}3\%$$

of the actual base rotation of the Mauvoisin Dam.

The settlement of the valley floor below the dam of about $u_r = 0{\cdot}6 \times 0{\cdot}0370 = 0{\cdot}0222$ m cannot be proven by *in situ* measurements because it is compensated by the vertical displacement (heave) of the crest due to the dam's vertical curvature[171]

$$u_x = w_{\mathrm{m}}^{\mathrm{W}}\left(\frac{16f_x}{3H}\right) = 0{\cdot}04946\left(\frac{16 \times 14{\cdot}3}{3 \times 237}\right) = 0{\cdot}0159\ \mathrm{m} \qquad (151)$$

and due to the effect of Poisson's ratio (after Equation (147)) of $u_\mu = 0{\cdot}0073$ m, together giving $u = 0{\cdot}0232$ m.

5.9.4. Heightening of the dam by 13·5 m

From 1990 to 1991 the Mauvoisin Dam was heightened by 13·5 m to a total of 250·5 m.[172] Because of this heightening the approximate analysis of Table 19 is supplemented as follows.

$$N_x^{\mathrm{G}} = -81{\cdot}96 - 0{\cdot}026\left[\left(\frac{14{\cdot}0 + 10{\cdot}8}{2}\right)13{\cdot}5 - 6{\cdot}0 \times 5{\cdot}0\right]$$
$$= -85{\cdot}40\ \mathrm{MN/m}$$

$$\sigma_x^{\mathrm{G}} = -\frac{85{\cdot}40}{53{\cdot}5}(1 \pm 1{\cdot}00) = \begin{matrix}-3{\cdot}19\\0\end{matrix}\ \mathrm{MN/m^2}$$

λ_x stays unchanged

$$M_x^{\mathrm{W}} = -0{\cdot}01(250{\cdot}5 - 49{\cdot}72) \times \frac{49{\cdot}72^2}{2} \times 0{\cdot}75$$
$$= -1861\ \mathrm{MN\ m/m}$$

$$N_x^{\mathrm{W}} = -14{\cdot}5 - 0{\cdot}01 \times 40{\cdot}0 \times 13{\cdot}5 = -19{\cdot}9\ \mathrm{MN/m}$$

$$\Delta M_x^{\mathrm{W}} = 19{\cdot}9 \times 18{\cdot}0 = 358\ \mathrm{MN\ m/m}$$

$$\sigma_x^{\mathrm{W}} = -\frac{19{\cdot}9}{53{\cdot}5} \pm \frac{6(1861 - 358)}{53{\cdot}5^2} = -0{\cdot}37 \pm 3{\cdot}15$$
$$= \begin{matrix}2{\cdot}78\\-3{\cdot}52\end{matrix}\ \mathrm{MN/m^2}$$

$$\sigma_x^{G+W} = \begin{matrix} -3{\cdot}19 & +2{\cdot}78 \\ 0 & -3{\cdot}52 \end{matrix} = \begin{matrix} -0{\cdot}41 \\ -3{\cdot}52 \end{matrix}\ \text{MN/m}^2$$

$$Q_x^{W} = 0{\cdot}01 \times 49{\cdot}72\left(250{\cdot}5 - \frac{49{\cdot}72}{2}\right)0{\cdot}75 = 84{\cdot}14\ \text{MN/m}$$

$$\tau_x = \frac{3}{2}\left(\frac{84{\cdot}14}{53{\cdot}5}\right) = 2{\cdot}36\ \text{MN/m}^2$$

$$\sigma_x = \frac{-0{\cdot}41 - 3{\cdot}52}{2} = -1{\cdot}97\ \text{MN/m}^2$$

$$\sigma_{1,2} = -\frac{1{\cdot}97}{2} \pm \left[\left(\frac{1{\cdot}97}{2}\right)^2 + 2{\cdot}36^2\right]^{1/2} = -0{\cdot}98 \pm 2{\cdot}56$$

$$= \begin{matrix} 1{\cdot}58 \\ -3{\cdot}54 \end{matrix}\ \text{MN/m}^2$$

$$p_y = 0{\cdot}01\left[\frac{250{\cdot}5}{2} - 49{\cdot}72 \times 0{\cdot}75 + \left(\frac{49{\cdot}72^2}{2 \times 250{\cdot}5}\right)0{\cdot}75\right]$$

$$= 0{\cdot}9166\ \text{MN/m}^2$$

λ_y practically unchanged

$$\sigma_{yA}^{W} = \frac{0{\cdot}9166}{0{\cdot}8488} \times \begin{vmatrix} 0{\cdot}55 \\ -7{\cdot}85 \end{vmatrix} = \begin{matrix} 0{\cdot}59 \\ -8{\cdot}48 \end{matrix}\ \text{MN/m}^2$$

$$\sigma_{yC}^{W} = \frac{0{\cdot}9166}{0{\cdot}8488} \times \begin{vmatrix} -6{\cdot}14 \\ -2{\cdot}38 \end{vmatrix} = \begin{matrix} -6{\cdot}63 \\ -2{\cdot}57 \end{matrix}\ \text{MN/m}^2$$

σ_x^{T} and σ_y^{T} stay unchanged

$$\sigma_{m}^{W} = \frac{0{\cdot}9166}{0{\cdot}8488} \times 49{\cdot}46 = 53{\cdot}41\ \text{mm}$$

w_m^T stays unchanged.

A comparison of these values with those before the heightening (Table 19) reveals that stresses and displacements increased by about 8%. In the 33 years before the heightening the compressive concrete strength had increased by at least 50% and the tensile strength by at least 25%. Therefore, the safety of concrete also increased, despite increased stresses:

(a) in comparison by at least $50 - 8 = 42\%$
(b) in tension by at least $25 - 8 = 17\%$.

This favourable result led in the meantime to the heightening of other arch dams.[173] Nevertheless, the fact that the strength of bedrock does not increase with time should not be overlooked. Increasing rock stresses thus lower the safety factor of the dam's foundation.

5.10. Numerical example: Gage Dam (Fig. 91)

This very thin constant radius arch dam was built in the French Massif Central in 1953.[174] Basically, the Gage Dam is a double size version ($H = 37{\cdot}6$ m) of the Stevenson Creek Test Dam in California. Although a test dam, it also served practical purposes. The Gage Dam was carefully observed during the first filling of the reservoir in summer 1954.[175] The deviations from predicted behaviour are due firstly to the underestimated degree of fixing in the sound gneiss bedrock (design assumption $E_r/E_c = 0{\cdot}25$ and actual value about 0·75) and secondly to cracks in the horizontal concreting joints.[176,177] The alarming growth of these cracks in the severe winter of 1962 – 1963 led Electricité de France, as owner of the Gage Dam, to order the construction of a second arch dam of usual wall thickness just upstream of the first (thin) one.

5.10.1. *Approximate analysis*

The corresponding values can be taken from Table 19.

5.10.2. *Comparison with* in situ *measurements immediately after the first filling*

After the first filling of the reservoir, vertical strains due to hydrostatic pressure of $\varepsilon_{xu} = 382 \times 10^{-6}$ on the upstream face and of $\varepsilon_{xd} = -238 \times 10^{-6}$ on the downstream face were measured at El. 978 m or 5·6 m above the dam's base.[175] The approximate analysis (Table 19) gives the bending stress at the base of $\sigma_x^W = \pm 3{\cdot}58$ MN/m^2

and, with a Young's modulus of $E_c = 18\ 000\ \mathrm{MN/m^2}$, the lower strains of $\varepsilon_x^W = \pm 199 \times 10^{-6}$. From the vertical bending stress in the dam's upper third-point (Fig. 79) of

$$\sigma_x^W = \mp \frac{0{\cdot}208 \times 3{\cdot}94 \times 6}{1{\cdot}30^2} = \mp 2{\cdot}91\ \mathrm{MN/m^2}$$

follow the corresponding strains of $\varepsilon_x^W = \mp 162 \cdot 10^{-6}$, while values of

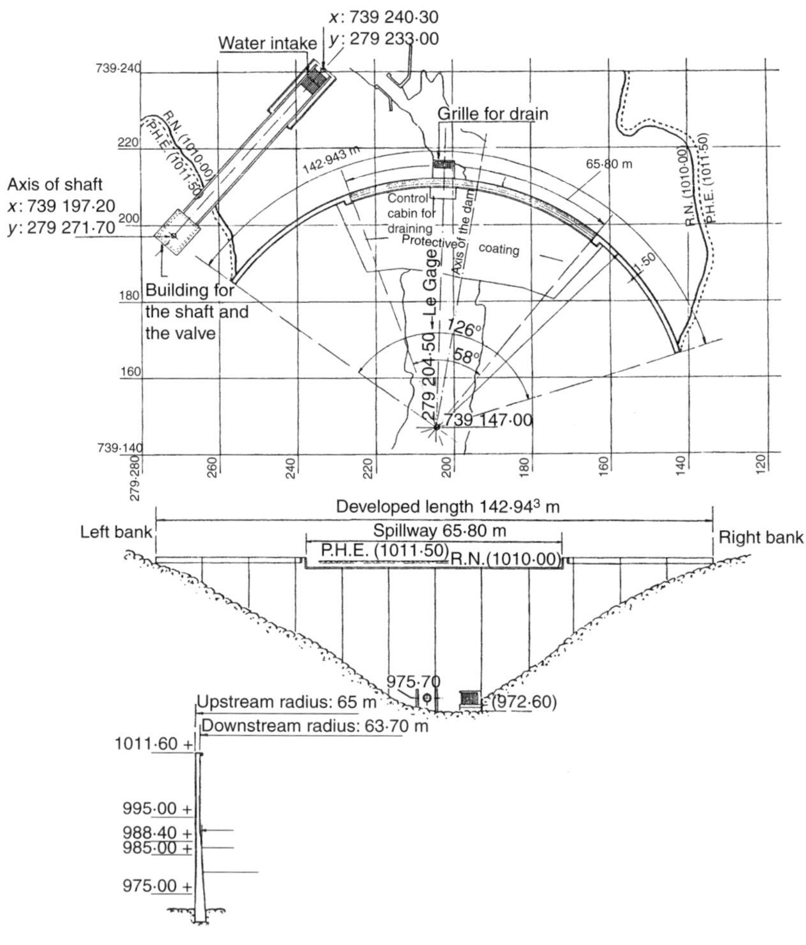

Fig. 91. Dimensions of the thin Gage constant radius arch dam[174]

$\varepsilon_{xu} = -53 \times 10^{-6}$ on the upstream face and $\varepsilon_{xd} = 168 \times 10^{-5}$ on the downstream face were measured at El. 994 m (Ref. 175, p. 63, Figs 1 and 2).

The measured horizontal strains at the crown of $\varepsilon_{yu} = -689 \times 10^{-6}$ on the upstream face and of $\varepsilon_{yd} = -464 \times 10^{-6}$ on the downstream face yield a horizontal direct stress of

$$\sigma_{yC} = -18\,000\left(\frac{689 + 464}{2}\right)10^{-6} = -10{\cdot}38\ \text{MN/m}^2$$

whilst the approximate analysis predicts only $\sigma_{yC}^{W} = -6{\cdot}88\ \text{MN/m}^2$. The horizontal bending moment at the crown follows from the measured strains as

$$M_{yC} = 18\,000(689 - 464)10^{-6}\left(\frac{1{\cdot}30^3}{12}\right) = 0{\cdot}74\ \text{MN m/m}$$

compared to a predicted value of $M_{yC}^{W} = 0$.

The agreement between radial displacements due to hydrostatic pressure is much better. For $E_c = 18\ 000\ \text{MN/m}^2$, the approximate analysis gives the predicted value of $w_m^W = 51{\cdot}95$ mm at midheight against the measured value of 58 mm.

5.10.3. *Comparison with later* in situ *measurements*

During the nine years of service of the first (thin) Gage Dam the following observations[57,176,177] were made.

The horizontal strain never surpassed the value of $\varepsilon_y^W = -450 \times 10^{-6}$ and the compressive stress of $\sigma_y^W = -8{\cdot}10\ \text{MN/m}^2$. The measured horizontal concrete stresses at El. 994 m of $\sigma_{yu}^W = -9{\cdot}8\ \text{MN/m}^2$ on the upstream face and of $\sigma_{yd}^W = -4{\cdot}7\ \text{MN/m}^2$ on the downstream face yield an average ring stress at the crown of

$$\sigma_{yC}^{W} = \frac{-9{\cdot}8 - 4{\cdot}7}{2} = -7{\cdot}25\ \text{MN/m}^2$$

(approximate analysis gives $-6{\cdot}88\ \text{MN/m}^2$) and the corresponding horizontal bending moment of

$$M_{yC}^{W} = (9{\cdot}8 - 4{\cdot}7)\left(\frac{1{\cdot}30^2}{6}\right) = 1{\cdot}44\ \text{MN m/m}$$

(approximate analysis gives a value of zero). The largest radial displacement at midheight was measured (Ref. 177, p. 1058, Fig. 4) to be $w_m^{W-T} = 75$ mm, whilst the approximate analysis yields a value of $w_m^{W-T} = 51{\cdot}95 + 22{\cdot}36 = 74{\cdot}31$ mm.

5.10.4. *Further observations*

The excessive formation of cracks, especially after the severe winter of 1962 to 1963, were most probably due to high bending stresses caused by the cooling down and subsequent warming up of the dam with an empty reservoir in the following summer 1963 which resulted in a change of sign of the surface stresses.

After the detection of an inclined shear crack in the dam's lower third (Fig. 92) dipping downstream in November 1963 and unsuccessful trials of waterproofing the upstream face in 1964 and 1965, the first (thin) Gage Dam was abandoned and replaced by a second arch dam of usual wall thickness immediately upstream of the first one (Fig. 93) in 1966.

5.10.5. *Upward sliding on the inclined valley flanks*

With a deadload of half the first (thin) Gage Dam by Simpson's rule of

$$G = 0{\cdot}025\left(\frac{37{\cdot}6}{3 \times 2}\right)(2{\cdot}57 \times 5{\cdot}0 + 4 \times 1{\cdot}40 \times 36{\cdot}7 + 1{\cdot}30 \times 71{\cdot}5)$$
$$= 48{\cdot}8 \text{ MN}$$

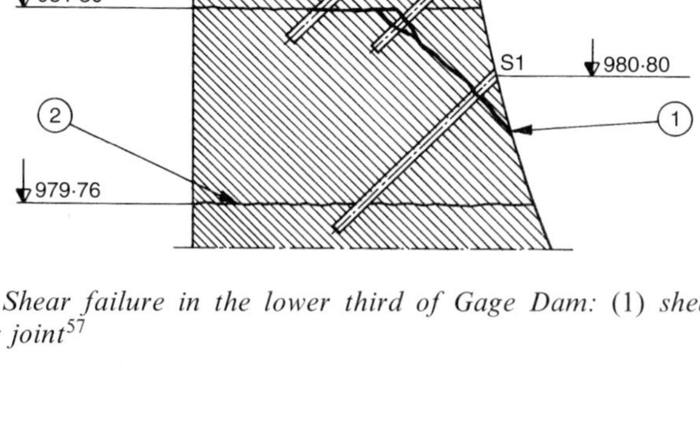

Fig. 92. Shear failure in the lower third of Gage Dam: (1) *shear crack and* (2) *concreting joint*[57]

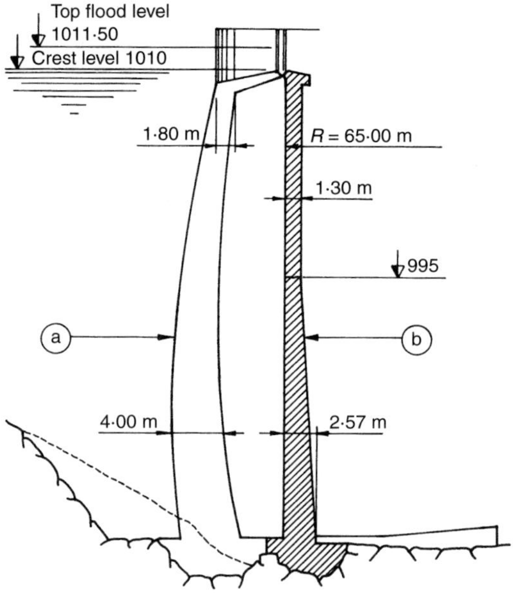

Fig. 93. Vertical section through first (b) *and second* (a) *Gage Dam*[57]

and the sum of the horizontal direct forces due to hydrostatic pressure at the crown given by Equation (133)

$$N_y = 0{\cdot}1482 \times 37{\cdot}6\left(64{\cdot}3 + \frac{1{\cdot}40}{2}\right) = -362{\cdot}1 \text{ MN}$$

the inclination of the resultant force is given by Equation (135) as

$$\tan\,\alpha = \frac{48{\cdot}8}{362{\cdot}1 \times 0{\cdot}853} = 0{\cdot}1580 \quad \text{and} \quad \alpha = 8^\circ\,59'$$

Then, the angle between the resultant force and the valley flank amounts after Equation (136) to $\varepsilon = 90^\circ - (9^\circ + 30^\circ) = 51^\circ$.

As *no* upward sliding was observed in the case of Gage Dam the effective angle of friction ϕ of this concrete dam upon the bedrock of gneiss was bigger than ε and the coefficient of friction $\tan\,\phi$ exceeded the value of $\tan\,\varepsilon = \tan\,51^\circ = 1{\cdot}235$.

5.11. Numerical example: Malpasset Dam (Fig. 94)

This 66·5 m high arch dam with sparing dimensions was built for the irrigation of the region of Fréjus in the French Department of Var from 1953 to 1955.[178,179] Because of the very irregular rainfall in this area of southern France it took four years to fill the reservoir for the first time. Heavy rains in November 1959 required the opening of the bottom outlet on 2 December 1959. Its capacity of only 40 m^3/s proved insufficient and five hours later the arch dam

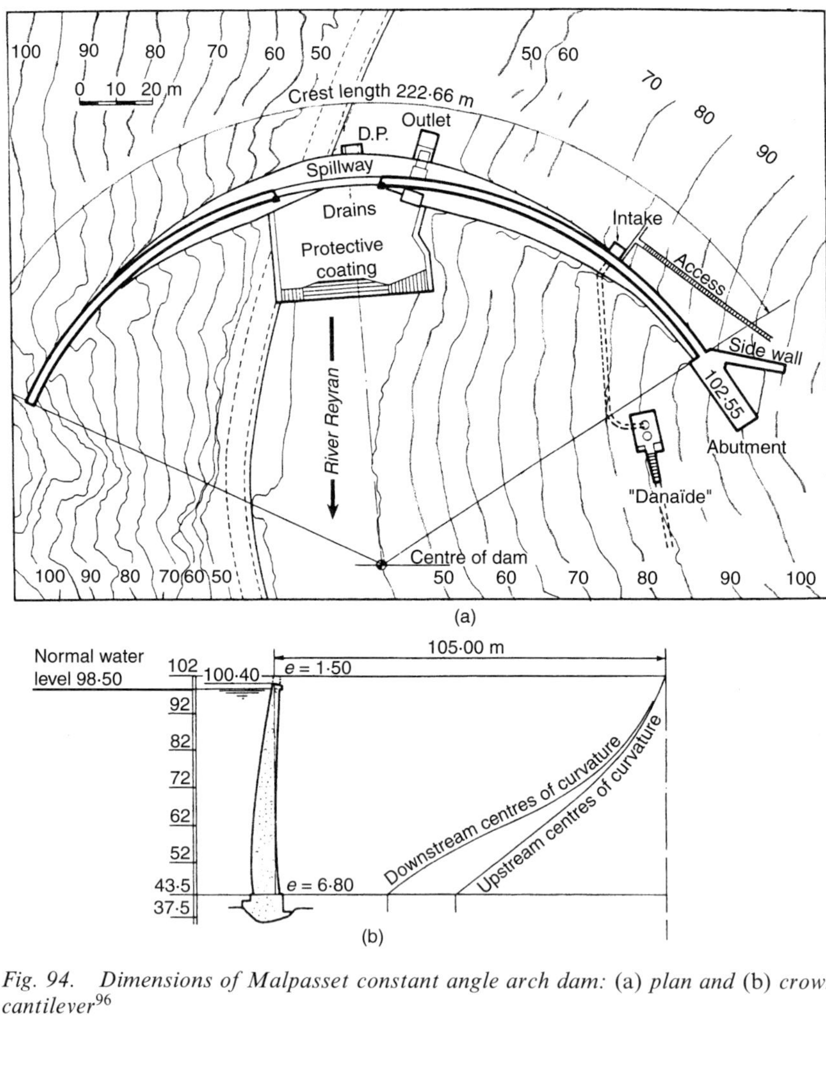

Fig. 94. Dimensions of Malpasset constant angle arch dam: (a) *plan and* (b) *crown cantilever*[96]

collapsed. The bedrock of gneiss below the left abutment was so compressed that it became impervious and the water pressure behind it rose to the full hydrostatic value (Fig. 95). This enormous load pushed away the left abutment with a portion of the bedrock upon a fault (Fig. 96) which had not been discovered previously. Twenty minutes after this event the 3 m high flood wave reached the town of Fréjus and destroyed it.

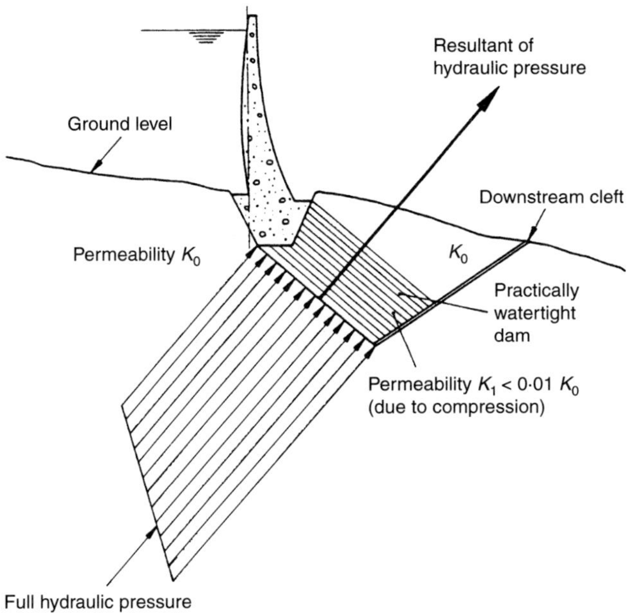

Fig. 95. Hydrostatic pressure acting on the left abutment of Malpasset Dam above the undetected fault[96]

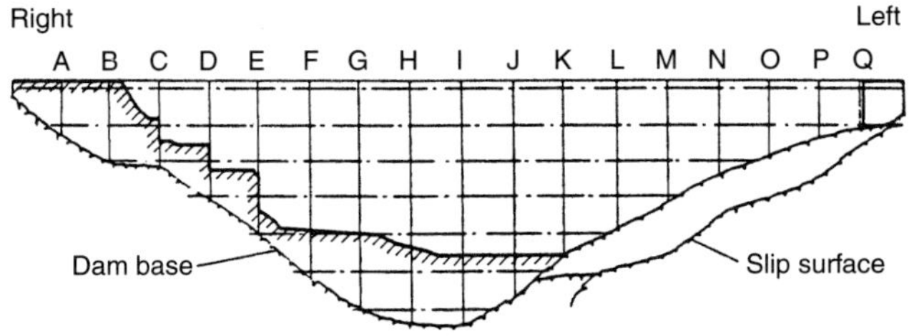

Fig. 96. Remnants of Malpasset Dam after failure[96]

5.11.1. Approximate analysis

For the corresponding values see Table 19.

5.11.2. Upward sliding on the inclined valley flanks

The deadload of half the Malpasset Dam (Fig. 94) by Simpson's rule of

$$G = 0{\cdot}025\left(\frac{66{\cdot}5}{3 \times 2}\right) \times (8{\cdot}53 \times 15{\cdot}0 + 4 \times 5{\cdot}63 \times 60{\cdot}6 + 1{\cdot}50 \times 111{\cdot}3) = 459\ \text{MN}$$

and the sum of the horizontal direct forces due to hydrostatic pressure at the crown after Equation (133) of

$$N_y = 0{\cdot}2074 \times 66{\cdot}5\left(80{\cdot}1 + \frac{5{\cdot}6}{2}\right) = -1143\ \text{MN}$$

give the inclination of the resultant force after Equation (135) as

$$\tan\ \alpha = \frac{459}{1143 \times 0{\cdot}859} = 0{\cdot}4675 \quad \text{and} \quad \alpha = 25^\circ\,03'$$

The angle between the resultant force and the inclined valley flank after Equation (136) of $\varepsilon = 90^\circ - (25^\circ + 30^\circ) = 35^\circ$ is considerably smaller than the angle of friction $\phi = 51^\circ$ of a concrete dam upon a bedrock of gneiss as found for the Gage Dam.

Contrary to other opinions[180] the Malpasset Dam was never endangered by upward sliding on the inclined valley flanks. The factor of safety amounted to at least

$$s = \frac{\tan\ \phi}{\tan\ \varepsilon} = \frac{1{\cdot}235}{0{\cdot}700} = 1{\cdot}76 \qquad (152)$$

5.12. Numerical example: Zeuzier Dam (Fig. 97)

This 156 m high constant angle arch dam in the asymmetrical Lienne Valley in the Swiss Canton of Valais was designed for a ring stress of (Equation (108))

$$\sigma_y^{\text{W}} = -\frac{0{\cdot}01 \times 156}{2 \times 15{\cdot}6}\left(83{\cdot}1 + \frac{15{\cdot}6}{2}\right) = -4{\cdot}55\ \text{MN/m}^2$$

at El. 1700 m and built from 1954 to 1956.[181] From the first filling of the reservoir in 1957 to the end of 1978 the Zeuzier Dam behaved completely normally (Ref. 137, pp. 213 – 222). Then extraordinary deformations (crest settlement of 125 mm, upstream radial displace-

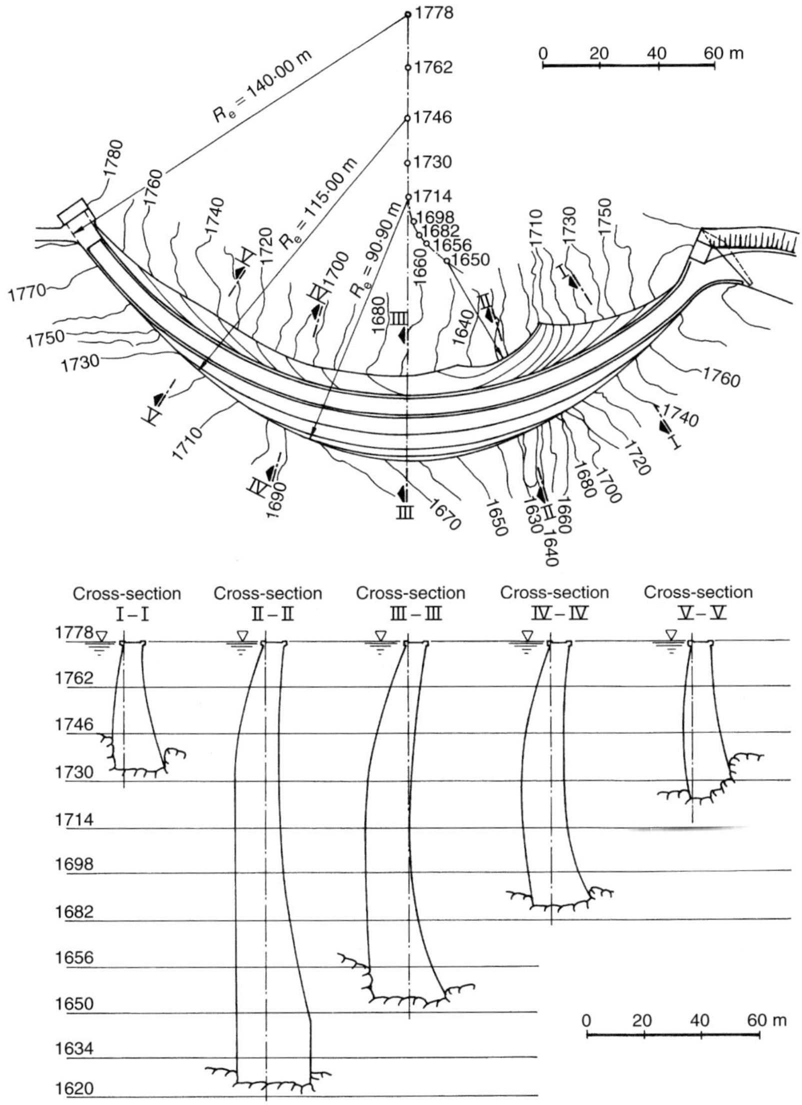

Fig. 97. Dimensions of unsymmetrical Zeuzier constant angle arch dam in Switzerland[184]

ment of −115 mm and a shortening of the abutment distance for the crest arch of 71 mm) were observed (Ref. 101, pp. 189 – 192). These were caused by the groundwater lowering in the pervious calcareous rock (malm formation) due to an exploration gallery for the planned Rawil road tunnel which was driven 140 m below the arch dam and acted as drainage.[182] A perimetral crack appeared on the dam's downstream face and five vertical construction joints opened on the upstream face in the crown region (Fig. 30). The originally fixed arches in the horizontal direction were transformed into three-hinged arches by the narrowing of the valley. To permit the upstream radial displacement of the dam, several parallel inclined cracks formed in the dam's base region (Fig. 31). After the reservoir had completely emptied and the crack formation had calmed down the different cracks were grouted with epoxy resins.[183] Thereafter, the reservoir was refilled in several steps. In 1988 the original maximum level at El. 1777 m was reached again.

5.12.1. Approximate analysis

Because of the valley's pronounced asymmetry the approximate analysis is based on the vertical section III (Fig. 97). The corresponding values can be taken from Table 19.

5.12.2. Comparison with model test

The radial displacement of the dam's vertical section III due to hydrostatic pressure was measured to be $w_{\mathrm{m}}^{\mathrm{W}} = 19{\cdot}3$ mm at El. 1718 m.[184] The approximate analysis, for $E_{\mathrm{c}} = 20\ 000$ MN/m^2 and $E_{\mathrm{r}}/E_{\mathrm{c}} = 0{\cdot}5$, yields the practically identical value of 19·54 mm. The vertical concrete stresses at the dam's base due to hydrostatic pressure were measured for vertical section III on the upstream face as $\sigma_{x\mathrm{u}}^{\mathrm{W}} = 2{\cdot}12$ MN/m^2 and on the downstream face as $\sigma_{x\mathrm{d}}^{\mathrm{W}} = -3{\cdot}68$ MN/m^2, whilst the approximate analysis predicts $\sigma_{x\mathrm{u}}^{\mathrm{W}} = 1{\cdot}80$ MN/m^2 and $\sigma_{x\mathrm{d}}^{\mathrm{W}} = -1{\cdot}98$ MN/m^2.

The horizontal concrete stresses due to hydrostatic pressure were measured at El. 1720 m and are detailed in Table 26.

The approximate analysis yields values for left and right abutments without distinction, as shown in Table 27.

Comparison of Tables 26 and 27 shows that there is a pronounced difference between measurement and prediction for values at the crown.

Table 26. Measured horizontal concrete stresses in Zeuzier Dam

	σ_y^W: MN/m²		
	Abutment		
Face of dam	Left	Right	Crown
Upstream	−0·68	−0·91	−4·12
Downstream	−4·35	−4·96	−0·19

Table 27. Predicted horizontal concrete stresses in Zeuzier Dam

	σ_y^W: MN/m²	
Face of dam	Abutment	Crown
Upstream	−0·55	−2·86
Downstream	−5·03	−2·72

5.12.3. *Comparison with* in situ *measurements*

Radial displacements at El. 1706·80 m due to hydrostatic pressure of the practically full reservoir in the vertical section III of 12 mm were measured on both 27 October 1961 and on 9 September 1962. As the dam is still warm at that time of the year due to the phase difference of air and concrete temperatures, the approximate analysis predicts the value of $w_m^W = 19{\cdot}54 - 4{\cdot}95 = 14{\cdot}59$ mm. The difference could be caused by an actually higher value of E_c.

5.13. Numerical example: Schiffenen Dam (Fig. 98)

This only 47 m high constant radius arch dam with a crest length of 417 m (ratio $B/H = 8{\cdot}87$) was built in the Sarine Valley of the Swiss Canton of Fribourg from 1961 to 1963. The bedrock is, as for the nearby Rossens Dam (Section 5.7) in the same valley, a marine sandstone called molasse with a low modulus of $E_r = 4000$ MN/m² as compared to the concrete's value of $E_c = 40\,000$ MN/m². It was only due to the low ratio of $E_r/E_c = 0{\cdot}1$ that the construction of this wide arch dam with a radius of curvature of $R_y = 200$ m became possible.[181]

5.13.1. *Approximate analysis*

For the corresponding values see again Table 19.

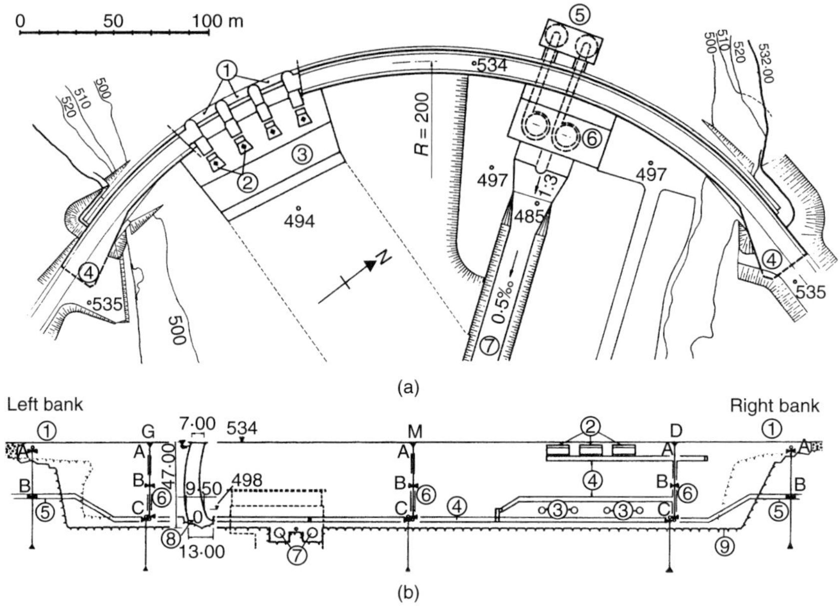

Fig. 98. Dimensions of Schiffenen constant radius arch dam in Switzerland: (a) *plan and* (b) *upstream view*[101]

5.13.2. Comparison with in situ *measurements*

The measured radial displacements at midheight (El. 514·8 m) amounted for a nearly full reservoir in 1983 (Fig. 99, after Ref. 101, p. 146, Figs. 4.5.1 – 7) in winter to 38 mm and in summer to 23 mm. The agreement with the approximate analysis of

$$w_{\mathrm{m}}^{\mathrm{W+T}} = 31{\cdot}02 \pm \frac{13{\cdot}80}{2} = \begin{matrix} 37{\cdot}92 \\ 24{\cdot}12 \end{matrix} \text{ mm}$$

is very good as long as the phase difference between air and concrete temperatures is taken into account.

5.14. Numerical example: Kölnbrein Dam (Fig. 100)

5.14.1 Damage and refit

This 200 m high constant angle arch dam was built from 1973 to 1977. Its design and construction have been described in depth.[185] The Kölnbrein Dam in the Malta Valley of the Austrian Province of Carinthia is situated in the central gneiss formation of the High

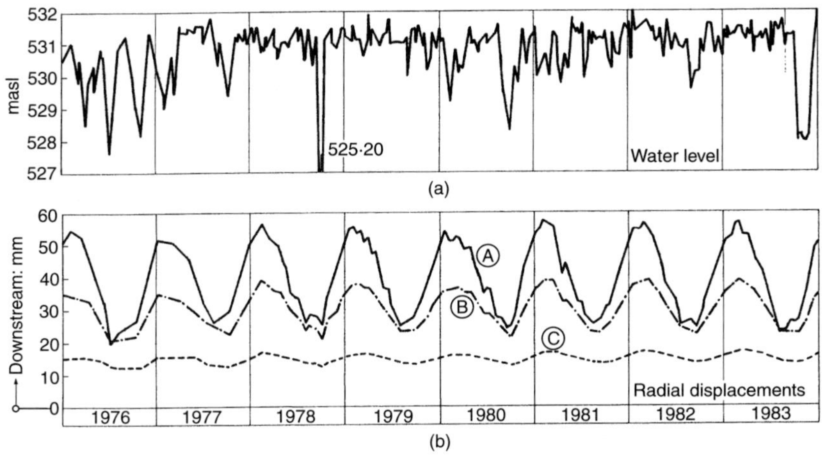

Fig. 99. Radial displacements of the crown cantilever from 1976 to 1983: (a) *reservoir level and* (b) *radial displacements. Curve A = crest at El. 533·2 m; B = 61% of full height at El. 514·8 m; C = 14% of height at El. 492·9 m)*[101]

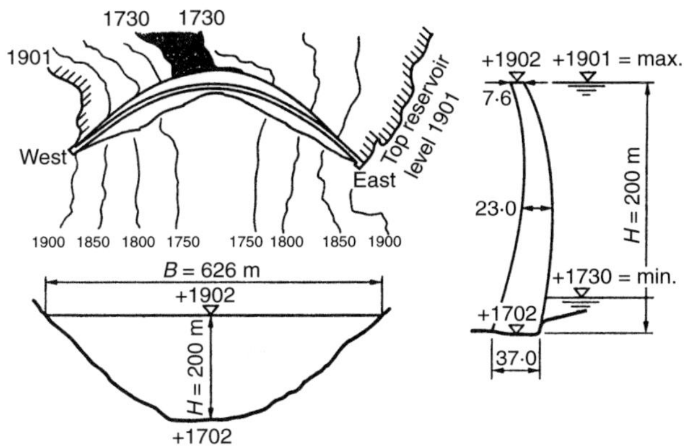

Fig. 100. Dimensions of original Kölnbrein Dam in Austria[185]

Tauern Mountains. Three different zones of gneiss must be distinguished. Massive gneiss dominates in the western (right) flank and the floor, bedded gneiss in the eastern (left) flank. At the toe of the left flank these two types of gneiss are separated by a zone of schistose gneiss (Fig. 32). The Kölnbrein Dam with a concrete volume of $1{\cdot}55 \times 10^6$ m^3 has a sophisticated shape. The horizontal sections vary from a hyperbola at the crest over a circle at midheight to a lying ellipse at the base.

Filling of the reservoir occurred in several stages. After surpassing El. 1860 m or 79% of the dam's height an unexpected seepage of 200 l/s and an uplift equal to the hydrostatic pressure were measured in summer 1978.[186] Subsequently, inclined cracks reaching from foundation level to the upstream face (Fig. 101) were noticed at the dam's foot and separating a concrete wedge at the dam's heel from the rest of it. After using the stored water for energy production in winter 1978 – 1979, a group of experts was called in April 1979. They took four years to reach a conclusion. In the meantime the owner and designer tried to overcome the problem with the help of groutings, drainages and a freeze curtain, but to no permanent avail. Later an apron with a new grout curtain was concreted upstream of the dam's heel to bridge cracks in the bedrock parallel to the dam (Fig. 101). In autumn 1983 the maximum water level was reached for the second time and a new inclined crack opened in the dam which started from the centre of the base to the upstream face. In spring 1984 the Austrian Water Authority restricted the impoundment level to El. 1880 m or 22 m below crest. By this measure the storage capacity of the reservoir was only 75% efficient.

The owner's Swiss consultant, *Lombardi*, classified the inclined cracks as 'shear failure'[85,187] and recommended the construction of a prop dam (Fig. 32(*c*)) at the downstream side. This equally arched prop dam was intended to take 22% of the hydrostatic load of 54 000 MN acting on the original Kölnbrein Dam and to reduce

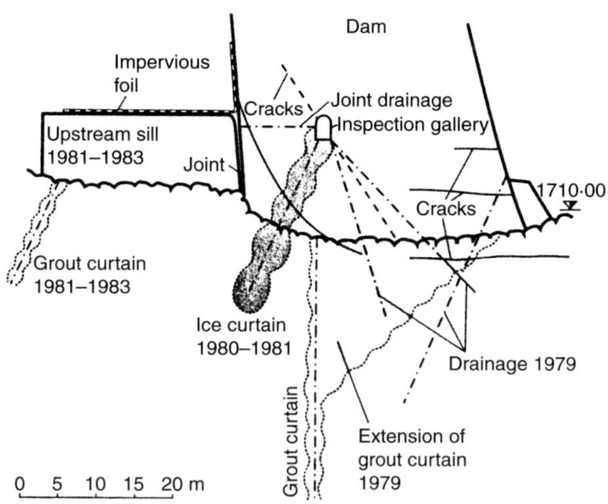

Fig. 101. Cracks and tentative refit measures at the base of Kölnbrein Dam[85]

the radial displacement at the base from 30 to 4 mm. The refit project was finally accepted by the Austrian Water Authority in December 1988 with 150 amendments and realized from 1989 to 1993. On 4 October 1993 the top water level at El. 1902 m was reached for the third time (in 15 years) and the values listed in Table 28 were measured.

5.14.2. *Approximate analysis of the original dam*

Kölnbrein Dam's sophisticated shape[185] with continuously varying thickness in the horizontal direction (abutment sections one and a half times thicker than the crown section) and low variability in the vertical direction (thickness at midheight 62% of base thickness) together with a comparatively thin base (ratio of base thickness to crest length $t_u/B = 37/626 = 0{\cdot}059$ or about 6% led to a high arch action with large radial displacements and high rock stresses.

The corresponding values of the approximate analysis can be taken from Table 19. The principal stresses must not be overlooked in thin arch dams.[156] The principal tensile stress of $\sigma_1 = 2{\cdot}28\ \mathrm{MN/m^2}$ is considerably larger than the tensile strength of the dam's concrete of about $1{\cdot}5\ \mathrm{MN/m^2}$. The predicted crack inclination of $\alpha = 33°$ coincides well with the *in situ* observation (Fig. 101). Due to the existing inclined cracks with an efficient base thickness of $t_u = 37/2 = 18{\cdot}5$ m, in 1983 the shear stress at the base increased to

$$\tau_x^{\mathrm{W}} = \frac{3}{2}\left(\frac{87{\cdot}5}{18{\cdot}5}\right) = 7{\cdot}09\ \mathrm{MN/m^2}$$

the normal stress increased to

$$\sigma_x^{\mathrm{G+W}} = \frac{-113{\cdot}8 - 6{\cdot}1}{18{\cdot}5} = -6{\cdot}48\ \mathrm{MN/m^2}$$

Table 28. Measurements of refitted Kölnbrein Dam

	1979	1983	1993
Maximum radial displacement: mm	150	132	104
Radial displacement at midheight: mm	130	113	84
Radial displacement at base: mm	30	25	9
Seepage: l/s	200	85	17
Water load transmitted to the prop dam: MN	0	0	9147

and the principal stresses increased to

$$\sigma_{1,2} = -\frac{6 \cdot 48}{2} \pm \left[\left(\frac{6 \cdot 48}{2} \right)^2 + 7 \cdot 09^2 \right]^{1/2}$$

$$= -3 \cdot 24 \pm 7 \cdot 80 = \begin{matrix} 4 \cdot 56 \\ -11 \cdot 04 \end{matrix} \ \mathrm{MN/m^2}$$

The already badly cracked Kölnbrein Dam was definitely endangered by this very high principal tensile stress.

The open (3 mm) horizontal cracks at the dam's toe showed that even the vertical tensile surface stress due to deadload of $\sigma_x^G = 1 \cdot 11\ \mathrm{MN/m^2}$ exceeded the reduced tensile strength (about $1 \cdot 0\ \mathrm{MN/m^2}$) of the horizontal concreting joints.

5.14.3. Comparison with *in situ* measurements

From the measured radial displacement at midheight for a full reservoir in 1979 (Fig. 102) of $w_m^{W+T} = 130$ mm and the predicted value of $w_m^{W+T} = 110 \cdot 00 - 9 \cdot 11 = 100 \cdot 89$ mm (because of the phase difference

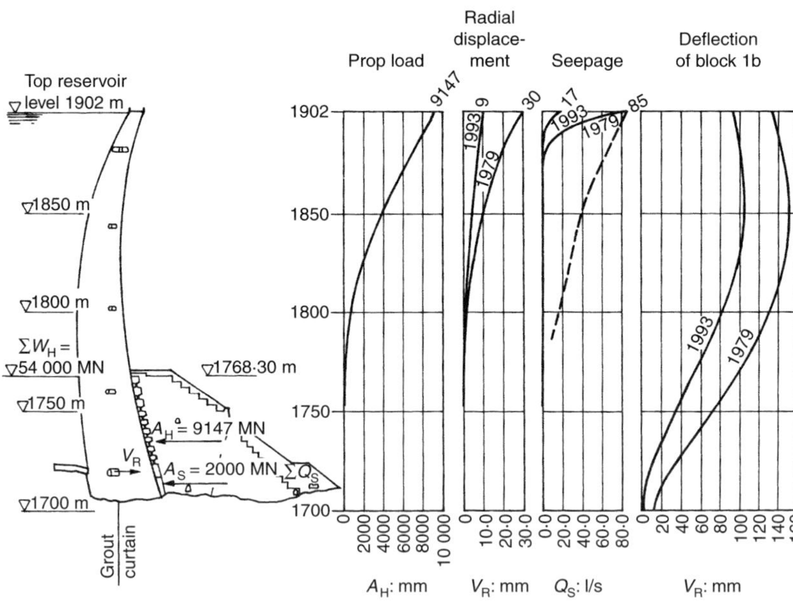

Fig. 102. Radial displacements of Kölnbrein Dam's block no. 16 at the first filling of the reservoir to top level in 1979, and at the third filling after refit and construction of prop dam in 1993[85]

between air and concrete temperatures, the dam was still warm in autumn) an indication of the reduced stiffness due to the inclined shear cracks at the dam's base can be obtained.

5.14.4. Action of the downstream prop dam

According to the measurements for a full reservoir on 4 October 1993 (Fig. 102) the prop dam took a load of 9147 MN or 16·94% of the total hydrostatic load on the Kölnbrein Dam. The corresponding shear force amounts to

$$\Delta Q_x^{\mathrm{W}} = \frac{3}{2}\left(\frac{9147}{338}\right) = 40{\cdot}6\ \mathrm{MN/m}$$

The remaining portion of hydrostatic pressure acting on the crown cantilever that has to be transmitted to the valley flanks is then given by

$$\begin{aligned} W_y &= \gamma_{\mathrm{w}}\frac{H^2}{2} - Q_x^{\mathrm{W}} - \Delta Q_x^{\mathrm{W}} = 200 - 87{\cdot}5 - 40{\cdot}6 \\ &= 71{\cdot}9\ \mathrm{MN/m} \end{aligned} \qquad (153)$$

Its average value (Equation (114))

$$p_y = \frac{71{\cdot}9}{200} = 0{\cdot}3595\ \mathrm{MN/m^2} < 0{\cdot}5625\ \mathrm{MN/m^2}$$

is, as expected, smaller than in the case without the prop dam. Therefore, the radial displacement at midheight due to hydrostatic pressure

$$w_{\mathrm{m}}^{\mathrm{W}} = \frac{0{\cdot}3595}{0{\cdot}5625} \times 110{\cdot}00 = 70{\cdot}30\ \mathrm{mm}$$

is equally reduced. As the dam was still warm at the time of measurement the predicted radial displacement at midheight amounts to

$$w_{\mathrm{m}}^{\mathrm{W+T}} = 70{\cdot}30 - 9{\cdot}11 = 61{\cdot}19\ \mathrm{mm}$$

This value is 27% smaller than the measured value of 84 mm because the existing stiffness of the refitted Kölnbrein Dam is smaller than that of the same dam free of cracks.

5.14.5. Stresses in the valley floor

Tensile stresses are caused in the valley floor by three different factors. First, the dam's deadload produces tensile stresses in the longitudinal direction of the valley immediately upstream and downstream of the dam (Fig. 103). From model tests[188] the surface stress is given by

$$\sigma_z^{G} = \frac{G}{3H} = \frac{113{\cdot}8}{3 \times 200} = 0{\cdot}19\ \mathrm{MN/m^2} \tag{154}$$

Second, the dam's shear force at the base due to hydrostatic pressure

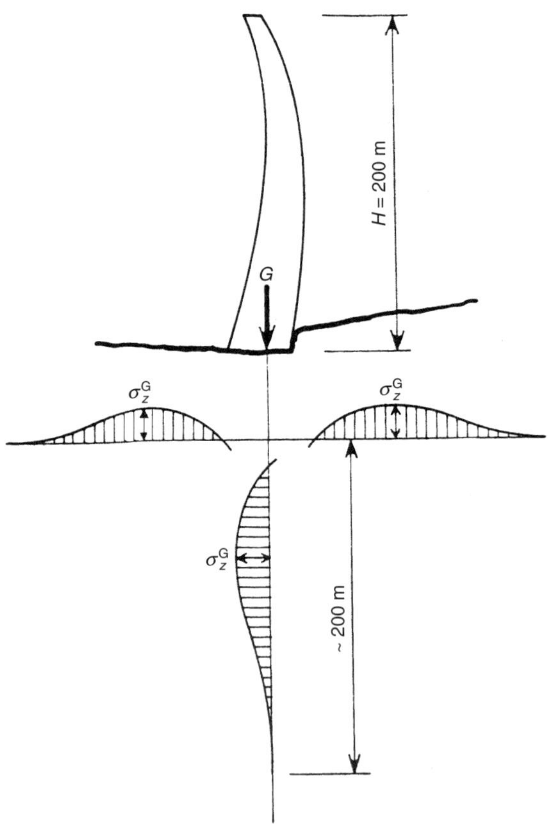

Fig. 103. Tensile stresses in the valley floor due to the dam's deadload[132]

(Fig. 104) produces a surface stress of

$$\sigma_z^{Q} = \frac{Q_x^{W}}{2t_u} = \frac{87\cdot5}{2\times37\cdot0} = 1\cdot18\ \text{MN/m}^2 \tag{155}$$

Third, the grout curtain below the dam (Fig. 105) produces a surface stress[88] of

$$\sigma_z^{W} = \frac{\gamma_w H}{2} = \frac{0\cdot01\times200}{2} = 1\cdot00\ \text{MN/m}^2 \tag{156}$$

The sum of these upstream tensile surface stresses of the valley floor before the start of the refit campaign amounted to

$$\sigma_z = \sigma_z^{G} + \sigma_z^{Q} + \sigma_z^{W} = 0\cdot19 + 1\cdot18 + 1\cdot00 = 2\cdot37\ \text{MN/m}^2 \tag{157}$$

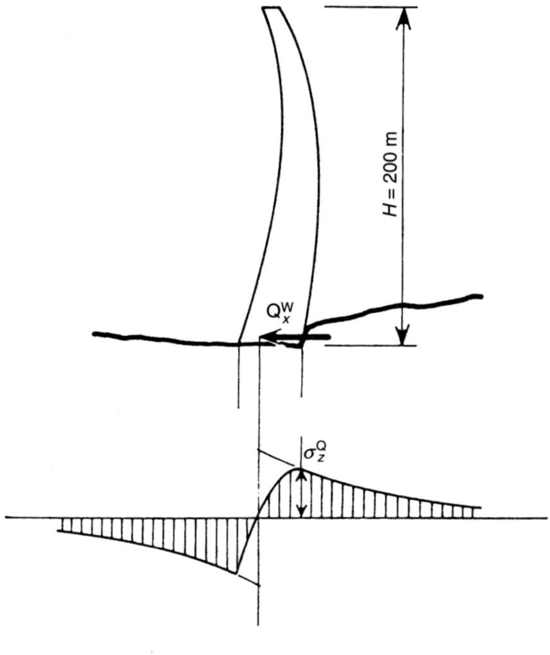

Fig. 104. Tensile stresses in the valley floor due to the shear force caused by hydrostatic pressure[132]

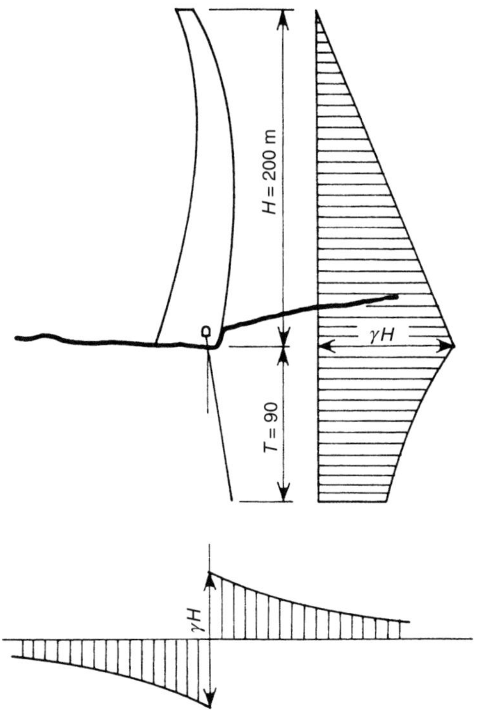

Fig. 105. Tensile stresses in the valley floor caused by hydrostatic pressure acting on the grout curtain[132]

This high value caused the opening of existing cracks and the formation of new ones in the valley's transverse direction. Because of these cracks the upstream concrete apron (Fig. 101) had to be built.

5.15. Rock stresses

It is worth noting that the actual rock stresses according to Equation (69) can be considerably larger than for a linear stress distribution, not only for massive gravity dams but also for arch dams.

6. Preliminary design examples

6.1. Embankment dams

The stability of a rockfill dam with a central vertical core (Fig. 106) has to be proven.

6.1.1. Construction settlements (Fig. 107)

For the shell:

- density: $\gamma_s = 19\ \text{kN/m}^3$ (rockfill)
- modulus of deformation: $E_s = 40\ \text{MN/m}^2$

For compressible underground:

- density: $\gamma_u = 24\ \text{kN/m}^3$ (valley floor)
- modulus of deformation: $E_u = 80\ \text{MN/m}^2$

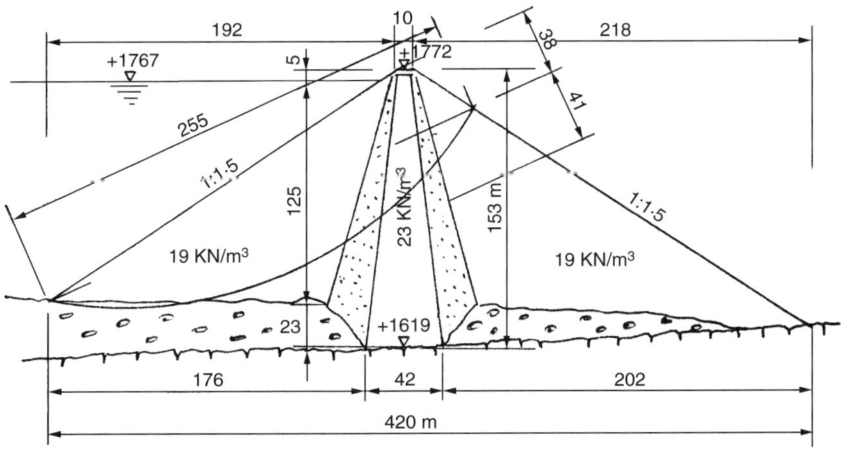

Fig. 106. Cross-section of embankment dam with fill densities (kN/m^3) and upstream slip circle

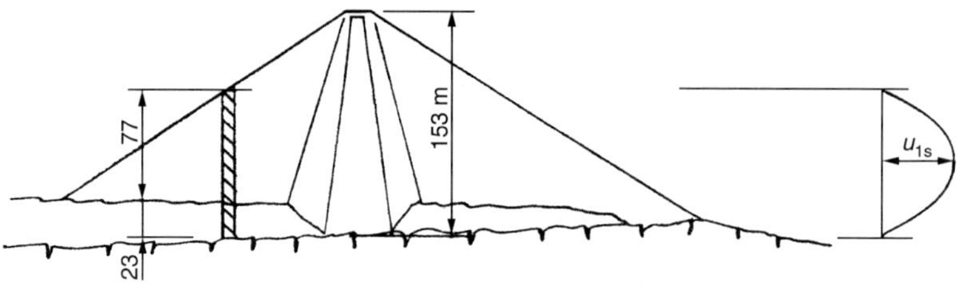

Fig. 107. Cross-section of embankment dam with idealized column used in the settlement analysis of the upsteam shell

- Weighted density of idealized column:

$$\gamma_m = \frac{\gamma_u t + \gamma_s h}{t + h} = \frac{24 \times 23 + 19 \times 77}{23 + 77} = \frac{2015}{100} = 20{\cdot}15 \text{ kN/m}^3$$
$$= 0{\cdot}02015 \text{ MN/m}^3$$

- Weighted modulus of deformation:

$$E_m = \frac{E_u t + E_s h}{t + h} = \frac{80 \times 23 + 40 \times 77}{23 + 77} = \frac{4920}{100} = 49{\cdot}2 \text{ MN/m}^2$$

- Maximum construction settlement at midheight of idealized column according to Equation (2):

$$u_{1s} = \frac{\gamma_m (t + h)^2}{4E_m} = \frac{0{\cdot}02015 \times 100^2}{4 \times 49{\cdot}2} = 1{\cdot}02 \text{ m}$$

6.1.2. Central core

Talus (upstream side enriched with addition of bentonite):

- density: $\gamma_c = 23$ kN/m^3 = 0·023 MN/m^3
- angle of internal friction: $\phi' = 35°$
- void ratio: $n = 18\%$
- coefficient of permeability: $k = 10^{-10}$ m/s
- modulus of deformation: $E_c = 150$ MN/m^2
- Maximum construction settlement at midheight of idealized column according to Equation (2):

$$u_{1s} = \frac{\gamma_c H^2}{4E_c} = \frac{0{\cdot}023 \times 153^2}{4 \times 150} = 0{\cdot}90 \text{ m}$$

6.1.3. Consolidation settlements

6.1.3.1. Upstream shell

- Maximum consolidation settlement of dam surface according to Equation (3):

$$u_{2s} = \frac{\gamma_m (t+h)^2}{2E_m} = 2u_{1s} = 2 \times 1{\cdot}02 = 2{\cdot}04 \text{ m}$$

6.1.3.2. Central core

- Maximum consolidation settlement of dam crest according to Equation (3):

$$u_{2s} = \frac{\gamma_c H^2}{2E_c} = 2u_{1c} = 2 \times 0{\cdot}90 = 1{\cdot}80 \text{ m}$$

6.1.4. Silo action

As the filters ($E_f = 250$ MN/m^2) settle less than the central core ($E_c = 150$ MN/m^2) the latter is propped by wall friction on the filters, such as stored material in a silo.

- Core width (at lower third of height): $b_c = 30$ m
- Horizontal earth pressure coefficient at rest: $\lambda_0 = 0{\cdot}45$
- Coefficient of wall friction: $\tan\delta = 0{\cdot}60$
- Bottom pressure of central core after *Jansen*'s classical silo theory according to Equation (4):

$$p_{cv} = \frac{\gamma_c b_c}{2\lambda_0 \tan\delta} = \frac{0{\cdot}023 \times 30}{2 \times 0{\cdot}45 \times 0{\cdot}60} = 1{\cdot}28 \text{ MN/m}^2$$

  (equalling 84% of the hydrostatic pressure at the dam base of $\gamma_w H = 0{\cdot}01 \times 153 = 1{\cdot}53$ MN/m^2)
- Overburden pressure: $p_v^0 = \gamma_c H = 0{\cdot}023 \times 153 = 3{\cdot}52$ MN/m^2
- Ratio of relief: $\eta = \dfrac{p_v^0 - p_{cv}}{p_v^0} = \dfrac{3{\cdot}52 - 1{\cdot}28}{3{\cdot}52} = 0{\cdot}64$

Due to the silo action the vertical filter stress is increased from the overburden pressure

$$p_v^0 = \gamma_f H = 0{\cdot}024 \times 153 = 3{\cdot}67 \text{ MN/m}^2 \ (= 100\%)$$

by the portion of wall friction

$$p_r = \eta\gamma_c\left(\frac{b_c}{2}\right) = 0{\cdot}64 \times 0{\cdot}023\left(\frac{30}{2}\right) = 0{\cdot}221\ \mathrm{MN/m^2}$$

divided by the base width of the filter $b_f = 20$ m

$$\Delta p_v = \frac{p_r H}{b_f} = \frac{0{\cdot}221 \times 153}{20} = 1{\cdot}69\ \mathrm{MN/m^2}$$

to the total of

$$p_v = p_v^0 + \Delta p_v = 3{\cdot}67 + 1{\cdot}69 = 5{\cdot}36\ \mathrm{MN/m^2}\ (= 146\%)$$

at the dam base.

6.1.5. Crest displacements

6.1.5.1. Hydrostatic pressure on the central core

- Average modulus of deformation (core, filter and downstream shell):

$$E_m = \frac{b_c E_c + b_f E_f + b_s E_s}{b_c + b_f + b_s} = \frac{0{\cdot}1 \times 150 + 0{\cdot}05 \times 250 + 0{\cdot}4 \times 40}{0{\cdot}1 + 0{\cdot}05 + 0{\cdot}4}$$
$$= \frac{43{\cdot}5}{0{\cdot}55} = 79{\cdot}1\ \mathrm{MN/m^2}$$

- Shear modulus according to Equation (50):

$$G_m = 0{\cdot}4E_m = 0{\cdot}4 \times 79{\cdot}1 = 31{\cdot}6\ \mathrm{MN/m^2}$$

- Crest displacement according to Equation (19):

$$w_1^W = \frac{\gamma_w h^2(h+f)}{4G_m B_y} = \frac{0{\cdot}01 \times 125^2 \times (125+5)}{4 \times 31{\cdot}6 \times 244} = 0{\cdot}66\ \mathrm{m}$$

6.1.5.2. Higher earth pressure of the downstream shell

- Densities of shells: downstream (dry) $\gamma_s = 19\ \mathrm{kN/m^3}$
  upstream (submerged) $\gamma_s' = 10\ \mathrm{kN/m^3}$

- Crest displacement according to Equation (21):

$$w_2^{\mathrm{W}} = -\frac{\lambda_0(\gamma_s - \gamma_s')H^3}{4G_m B_y} = -\frac{0{\cdot}45 \times 0{\cdot}009 \times 130^3}{4 \times 31{\cdot}6 \times 244} = -0{\cdot}29 \text{ m}$$

6.1.5.3. *Compression of valley floor material above bedrock by the reservoir's water load*

- Crest displacement according to Equations (23) and (24):

$$w_3^{\mathrm{W}} = -\frac{\gamma_w(H-f)T}{E_u}\frac{H}{B} = -\frac{0{\cdot}01 \times 125 \times 23}{80} \times \frac{130}{420}$$
$$= -0{\cdot}11 \text{ m}$$

6.1.5.4. *Total crest displacement*

• Hydrostatic pressure:	0·66 m
• Difference of earth pressures:	−0·29 m
• Reservoir load on valley floor:	−0·11 m
• Total towards downstream:	$w^{\mathrm{W}} = 0{\cdot}26$ m

6.1.6. *Slip safeties*

6.1.6.1. *Dam sliding bodily on the valley floor fill* (Fig. 108)

- Hydrostatic load according to Equation (18):

$$W = \frac{\gamma_w}{2}(H-f)^2 = \frac{0{\cdot}01 \cdot 125^2}{2} = 78{\cdot}1 \text{ MN/m}$$

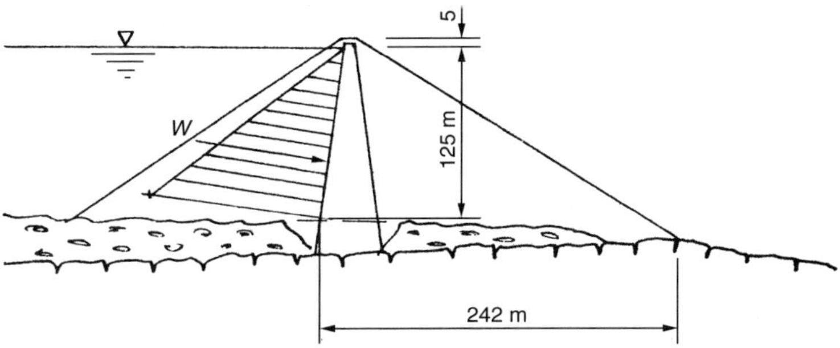

Fig. 108. Dam trying to slide bodily on the valley floor fill

- Average shear stress according to Equation (29):

$$\tau = \frac{W}{B_y} = \frac{78\cdot 1}{242} = 0\cdot 323\ \text{MN/m}^2$$

- Average vertical stress (core, filter and shell) according to Equation (30)

$$\sigma = \frac{\gamma_c b_c + \gamma_f b_f + \gamma_s b_s}{b_c + b_f + b_s}\left(\frac{H}{2}\right)$$
$$= \frac{0\cdot 023 \times 38 + 0\cdot 024 \times 19 + 0\cdot 019 \times 185}{38 + 19 + 185}\left(\frac{130}{2}\right)$$
$$= \frac{4\cdot 845}{242} \times 65 = 1\cdot 301\ \text{MN/m}^2$$

- Condition of sliding according to Equation (31) not fulfilled:

$$\frac{\tau}{\sigma} = \frac{0\cdot 323}{1\cdot 301} = 0\cdot 248 < \tan\phi' = 0\cdot 70$$

- Factor of safety:

$$s = \frac{0\cdot 70}{0\cdot 248} = 2\cdot 82 > 2\cdot 0$$

6.1.6.2. Upstream slope for full reservoir

The critical loading 'rapid drawdown' is not relevant in this case because of the shell's high void ratio ($n = 29\%$).

- Vertical load of the slip body (Fig. 109) according to Equation (32):

$$G_D - U = \gamma'\left(\frac{a_1}{2} + \frac{2}{3}a_2\right)b = 0\cdot 01\left(\frac{38}{2} + \frac{2}{3} \times 41\right)255$$
$$= 118\cdot 1\ \text{MN/m}$$

- Average slope of the slip circle (Fig. 109): $\alpha = 25°$
- Normal force component according to Equation (33*a*):

$$N = (G_D - U)\cos\alpha = 118\cdot 1 \times 0\cdot 9063 = 107\cdot 0\ \text{MN/m}$$

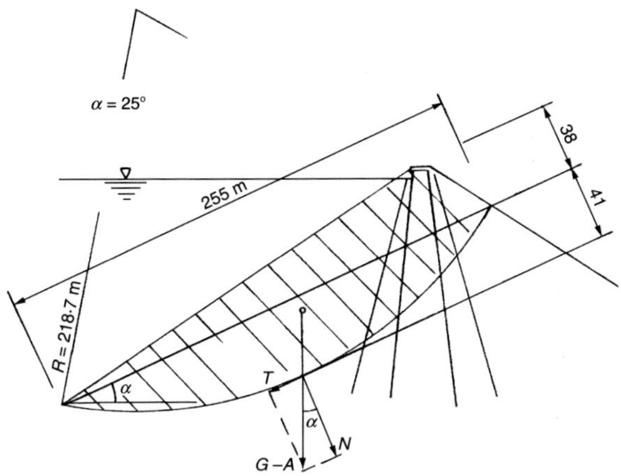

Fig. 109. Slip body of the dam's upstream side subjected to its submerged weight

- Tangential force component according to Equation (33*b*):

  $T = (G_D - U)\sin\ \alpha = 118{\cdot}1 \times 0{\cdot}4226 = 49{\cdot}9\ \text{MN/m}$

- Average shear strength of the dam:

$$\phi'_m = \frac{43^0 \times 244 + 35^0 \times 32}{244 + 32} = \frac{11\,612^0}{276}$$
$$= 42{\cdot}07^0 \quad \text{and} \quad \tan\ \phi'_m = 0{\cdot}9027$$

- Factor of safety according to Equation (34):

$$s = \frac{N}{T}\tan\ \phi'_m = \frac{107{\cdot}0}{49{\cdot}9} \times 0{\cdot}9027 = 1{\cdot}94 > 1{\cdot}5$$

6.1.7. Earthquakes

- Horizontal component of the dam's response acceleration

  $a_{Eh} = 0{\cdot}1g$

6.1.7.1. Reservoir empty

Slip safety of the dam's upstream side (Fig. 110).

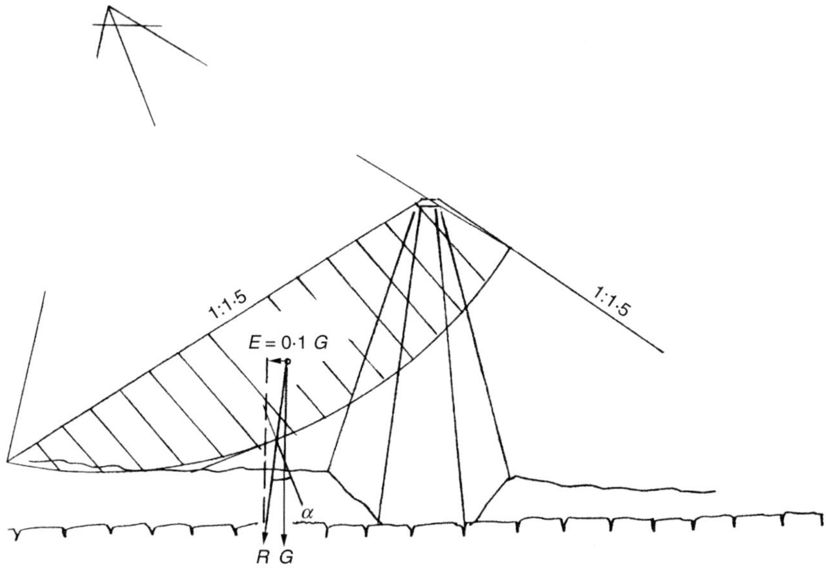

Fig. 110. Slip body of the dam's upstream side subjected to its dry weight and the horizontal earthquake response acceleration

- Weight of the dry dam body according to Equation (32):

$$G_D = \gamma_m\left(\frac{a_1}{2} + \frac{2}{3}a_2\right)b = 0{\cdot}020\left(\frac{38}{2} + \frac{2}{3} \times 41\right)255 = 236{\cdot}2\ \text{MN/m}$$

- Horizontal earthquake response load according to Equation (35*a*):

$$E_h = 0{\cdot}1G_D = 0{\cdot}1 \times 236{\cdot}2 = 23{\cdot}6\ \text{MN/m}$$

- Resulting load:

$$R = \left(G_D^2 + E_h^2\right)^{1/2} = (236{\cdot}2^2 + 23{\cdot}6^2)^{1/2} = 237{\cdot}4\ \text{MN/m}$$

- Angle between resulting load and slip circle:

$$\alpha = 25^\circ + 5^\circ\,43' = 30^\circ\,43'$$

- Normal force component according to Equation (33*a*):

$$N = R\cos\,\alpha = 237{\cdot}4 \times 0{\cdot}8597 = 204{\cdot}2\ \text{MN/m}$$

- Tangential force component according to Equation (33*b*):

  $T = R \sin \alpha = 237{\cdot}4 \times 0{\cdot}5108 = 121{\cdot}3$ MN/m

- Factor of safety according to Equation (34)

  $$s = \frac{N}{T} \tan \phi_m = \frac{204{\cdot}1}{121{\cdot}3} \times 0{\cdot}9027 = 1{\cdot}86 > 1{\cdot}5$$

6.1.7.2. Reservoir full

Slip safety of the upstream side: the earthquake response acceleration acts on both the dam body and the participating body of water above the upstream shell.

- $W = \gamma_w \left(\frac{bh}{2}\right) a_w = 0{\cdot}01 \left(\frac{183 \times 121}{2}\right) 61 = 110{\cdot}7 \times 61$
  $= 6753$ MNm/m
- G_D (the weight of the submerged dam body and the water equals approximately the weight of the mosit dam body; see Section 6.1.7.1. Reservoir empty) $= 236{\cdot}2 \times 129 = 30\ 470$ MN m/m

---

- Vertical load: $V = 346{\cdot}9$ MN/m
- Total moment $M = 37\ 223$ MN m/m
- Lever arm: $a_V = \frac{37\,223}{346{\cdot}9} = 107{\cdot}3$ m
- $0{\cdot}1W = 0{\cdot}1 \times 110{\cdot}7 = 11{\cdot}07 \times 40{\cdot}3 = 446{\cdot}1$ MN m/m
- $0{\cdot}1G_D = 0{\cdot}1 \times 236{\cdot}2 = 23{\cdot}62 \times 70{\cdot}0 = 1653{\cdot}4$ MN m/m

---

- Horizontal load: $H = 34{\cdot}69$ MN/m

  Total moment $M = 2099{\cdot}5$ MN m/m
- Lever arm: $a_H - \frac{2099{\cdot}5}{34{\cdot}69} = 60{\cdot}5$ m
- Resulting load:

  $R = (V^2 + H^2)^{1/2} = (346{\cdot}9^2 + 34{\cdot}69^2)^{1/2} = 348{\cdot}6$ MN/m

- Angle between resulting load and slip circle (graphically):

  $\alpha = 20^\circ\ 30'$

- Normal load component according to Equation (33*a*):

  $N = R \cos \alpha = 348{\cdot}6 \times 0{\cdot}9367 = 326{\cdot}5$ MN/m

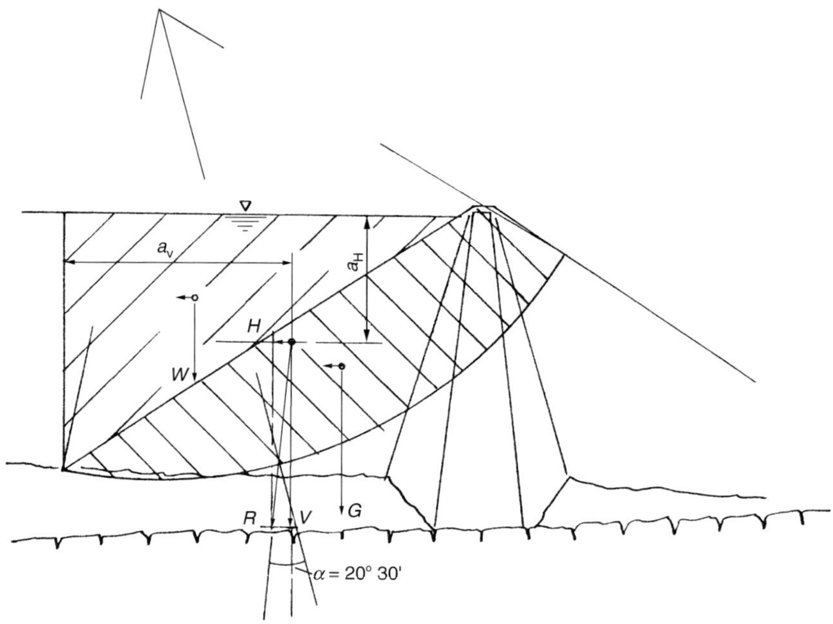

Fig. 111. Slip body of the dam's upstream side subjected to its submerged weight, the participating body of water and the horizontal earthquake response acceleration of both

- Tangential load component according to Equation (33*b*):

  $T = R \sin \alpha = 348.6 \times 0{\cdot}3502 = 122{\cdot}1$ MN/m

- Factor of safety according to Equation (34):

$$s = \frac{N}{T} \cos \phi'_{\mathrm{m}} = \frac{326{\cdot}5}{122{\cdot}1} \times 0{\cdot}9027 = 2{\cdot}41 > 1{\cdot}5$$

6.1.8. Conclusion

The embankment dam as designed is sufficiently safe.

6.2. Gravity dams

Only a 1 m thick vertical slice of the concrete dam (Fig. 112) is analysed. Due to the elasticity of the bedrock ($E_{\mathrm{r}} = 15\ 000$ MN/m^2) the extreme contact stresses at the dam's base are about 40% higher

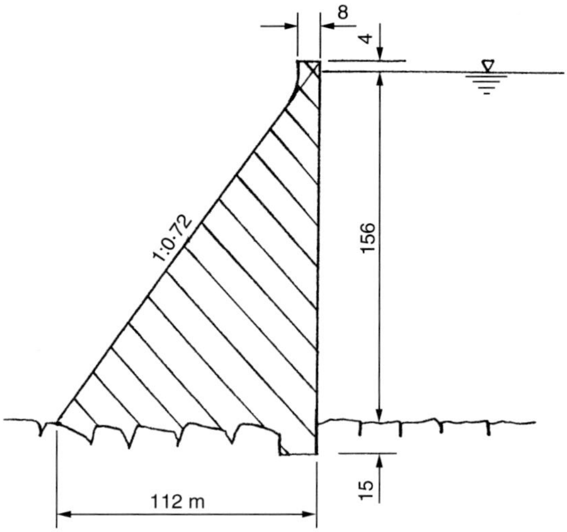

Fig. 112. Cross-section of gravity dam

than the calculated values based on the assumption of a linear stress distribution.

6.2.1. Stresses

- Density of concrete: $\gamma_c = 24\ \text{kN/m}^3 = 0{\cdot}024\ \text{MN/m}^3$
- Strength of concrete C35: $R_c = 0{\cdot}6 \times 35 = 23\ \text{MN/m}^2$
- Modulus of elasticity: $E_c = 20\ 000\ \text{MN/m}^2$

6.2.1.1. *Deadload* (Fig. 113)

For a triangular cross-section of the dam

- $G = \gamma_c \dfrac{BH}{2} = 0{\cdot}024\left(\dfrac{112 \times 160}{2}\right) = 215{\cdot}0\ \text{MN/m}$
- contact stresses (linear distribution):
  upstream $\sigma_u^G = -0{\cdot}024 \times 160 = -3{\cdot}84\ \text{MN/m}^2$
  downstream $\sigma_d^G = 0$

6.2.1.2. *Deadload, hydrostatic pressure and uplift* (Fig. 114)

- Hydrostatic pressure: $p_w = \gamma_w h = 0{\cdot}01 \times 156 = 1{\cdot}56\ \text{MN/m}^2$
- Hydrostatic load: $W = p_w\left(\dfrac{h}{2}\right) = 1{\cdot}56\left(\dfrac{156}{2}\right) = 121{\cdot}7\ \text{MN/m}$

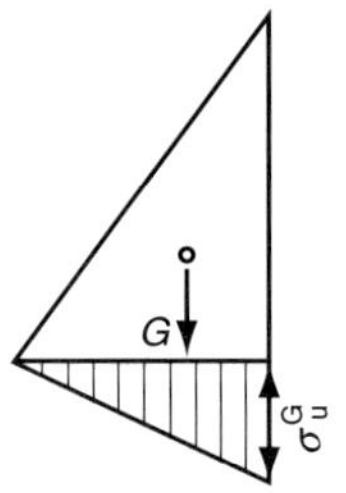

Fig. 113. Contact stresses at the dam base due to deadload only

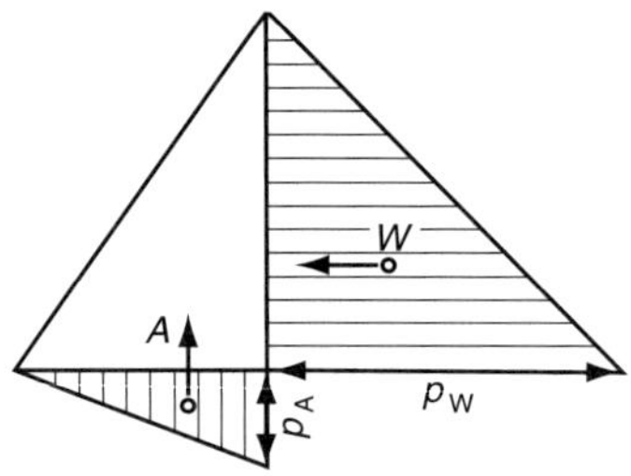

Fig. 114. Cross-section of gravity dam subjected to hydrostatic pressure and uplift

- Uplift pressure: $p_\mathrm{u} = \dfrac{p_\mathrm{W}}{4} = \dfrac{1{\cdot}56}{4} = 0{\cdot}39\ \mathrm{MN/m^2}$
- Uplift load: $U = p_\mathrm{u}\left(\dfrac{B}{2}\right) = 0{\cdot}39\left(\dfrac{112}{2}\right) = 21{\cdot}8\ \mathrm{MN/m}$
- Distance of the resulting force from the upstream third-point of the dam base:

$$c^{\mathrm{G-U+W}} = \frac{h}{3}\left(\frac{W}{G-U}\right) = \frac{156}{3}\left(\frac{121{\cdot}7}{215{\cdot}0 - 21{\cdot}8}\right)$$
$$= 32{\cdot}76\ \mathrm{m} < \frac{B}{3} = \frac{112}{3} = 37{\cdot}33\ \mathrm{m}$$

- Contact stresses (linear distribution):

$$\sigma^{\mathrm{G-U+W}} = \frac{G-U}{B}\left(1 \pm \frac{6(c - B/6)}{B}\right)$$

- Downstream side:

$$\sigma_{\mathrm{u}}^{\mathrm{G-U+W}} = \frac{215{\cdot}0 - 21{\cdot}8}{112}\left(1 + \frac{6(32{\cdot}76 - 112/6)}{112}\right)$$
$$= -1{\cdot}725\left(1 + \frac{14{\cdot}09}{18{\cdot}67}\right) = -3{\cdot}03\ \mathrm{MN/m^2}$$

- Upstream side:

$$\sigma_{\mathrm{u}}^{\mathrm{G-U+W}} = -1{\cdot}725(1 - 0{\cdot}755) = -0{\cdot}42\ \mathrm{MN/m^2}$$

- Both values are compressive stresses.

6.2.1.3. Earthquake with empty reservoir

- Horizontal response acceleration of the dam due to the earthquake: $a_{\mathrm{Eh}} = 0{\cdot}1g$
- Maximum horizontal earthquake pressure at dam base:

$$p_{\mathrm{E}} = \left(\frac{a_{\mathrm{Eh}}}{g}\right)\gamma_{\mathrm{c}}^{\mathrm{B}} = 0{\cdot}1 \times 0{\cdot}024 \times 112 = 0{\cdot}269\ \mathrm{MN/m^2}$$

- Horizontal earthquake load (Fig. 115):

$$E = p_{\mathrm{E}}\left(\frac{H}{2}\right) = 0{\cdot}269\left(\frac{160}{2}\right) = 21{\cdot}5\ \mathrm{MN/m}$$

- Eccentricity of the resulting force:

$$e^{\mathrm{G+E}} = \frac{H}{3}\left(\frac{E}{G}\right) = \frac{160}{3}\left(\frac{21{\cdot}5}{215{\cdot}0}\right) = 5{\cdot}33\ \mathrm{m}$$

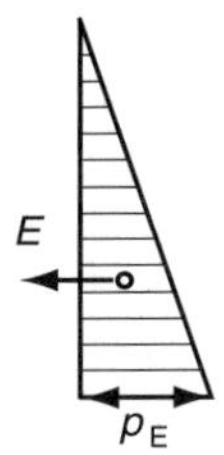

Fig. 115. Idealized earthquake response load

- Distance of the resulting force from the dam's heel:

$$a = \frac{B}{3} - c^{\mathrm{G+E}} = \frac{112}{3} - 5{\cdot}33 = 32{\cdot}00 \text{ m}$$

- Contact stress (linear distribution) without tension in case of earthquake directed upstream (Fig. 116):

$$\sigma_{\mathrm{u}}^{\mathrm{G+E}} = \frac{2G}{3a} = \frac{2 \times 215{\cdot}0}{3 \times 32{\cdot}00} = -4{\cdot}48 \text{ MN/m}^2$$

- Contact stress in the case of earthquake directed downstream (Fig. 117):

$$\sigma_{\mathrm{d}}^{\mathrm{G+E}} = \frac{G}{B}\left(1 \pm \frac{6(B/6 - c)}{B}\right)$$

○ Downstream side:

$$\sigma_{\mathrm{d}}^{\mathrm{G+E}} = -\frac{215{\cdot}0}{112}\left(1 - \frac{6(18{\cdot}67 - 5{\cdot}33)}{112}\right)$$
$$= -1{\cdot}92\left(1 - \frac{13{\cdot}34}{18{\cdot}67}\right) = -0{\cdot}55 \text{ MN/m}^2$$

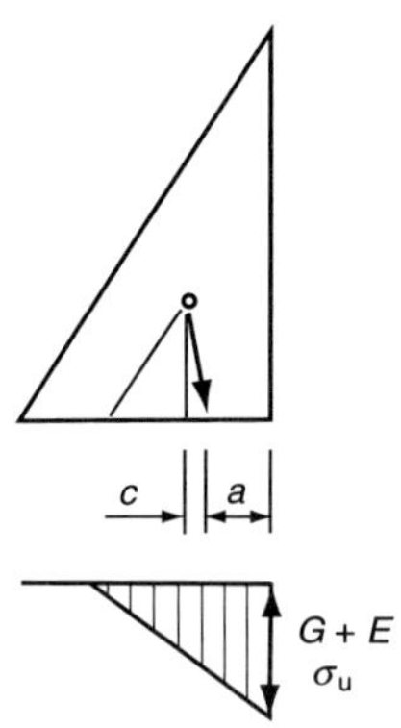

Fig. 116. Contact stresses at the dam base without tension in case of an earthquake directed upstream (reservoir empty)

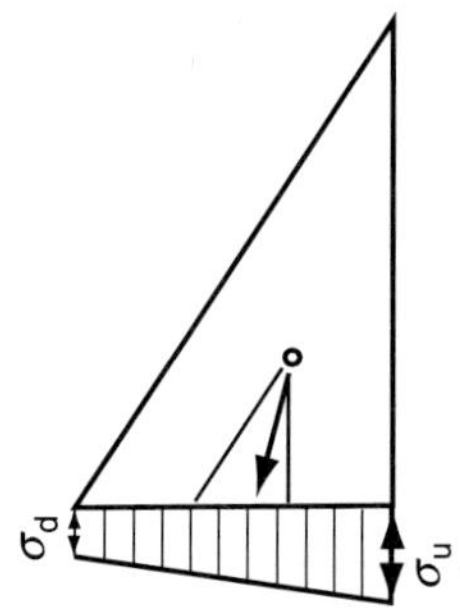

Fig. 117. Contact stresses at the dam base in case of an earthquake directed downstream (reservoir empty)

- Upstream side:

$$\sigma_{\mathrm{u}}^{\mathrm{G+E}} = -1{\cdot}92(1 + 0{\cdot}715) = -3{\cdot}29\ \mathrm{MN/m^2}.$$

6.2.1.4. Earthquake with full reservoir

- Horizontal earthquake pressure at dam base:

$$p_{\mathrm{E}} = \frac{a_{\mathrm{Eh}}}{g}(\gamma_{\mathrm{c}}B + \gamma_{\mathrm{w}}h) = 0{\cdot}1(0{\cdot}024 \times 112 + 0{\cdot}01 \times 156)$$
$$= 0{\cdot}425\ \mathrm{MN/m^2}$$

- Horizontal earthquake load:

$$E = p_{\mathrm{E}}\left(\frac{H}{2}\right) = 0{\cdot}425\left(\frac{160}{2}\right) = 34{\cdot}0\ \mathrm{MN/m}$$

- Eccentricity of the resulting force (E directed downstream):

$$c^{\mathrm{G-U+W+E}} = \frac{H}{3}\left(\frac{W+E}{G-U}\right) = \frac{160}{3}\left(\frac{121{\cdot}7 + 34{\cdot}0}{215{\cdot}0 - 21{\cdot}8}\right)$$
$$= 42{\cdot}98\ \mathrm{m} > \frac{B}{3} = 37{\cdot}33\ \mathrm{m}$$

- Distance of the resulting force from the dam's toe:

$$a = \tfrac{2}{3}B - c^{\mathrm{G-U+W+E}} = \tfrac{2}{3}112 - 42{\cdot}98 = 31{\cdot}69\ \mathrm{m}$$

- Contact stress without tension in the case of earthquake directed downstream (Fig. 118):

$$\sigma_d^{G-U+W+E} = \frac{2(G-U)}{3a} = \frac{2(215{\cdot}0 - 21{\cdot}8)}{3 \times 31{\cdot}69} = -4{\cdot}06 \text{ MN/m}^2$$

- Eccentricity of the resulting force (E directed upstream):

$$c^{G-U+W-E} = \frac{H}{3}\left(\frac{W-E}{G-U}\right) = \frac{160}{3}\left(\frac{121{\cdot}7 - 34{\cdot}0}{215{\cdot}0 - 21{\cdot}8}\right)$$
$$= 24{\cdot}21 \text{ m} < B/3 = 37{\cdot}33 \text{ m}$$

- Contact stresses (linear distribution) in case of earthquake directed upstream (Fig. 119):

$$\sigma^{G-U+W-E} = \frac{G-U}{B}\left(1 \pm \frac{6(c - B/6)}{B}\right)$$

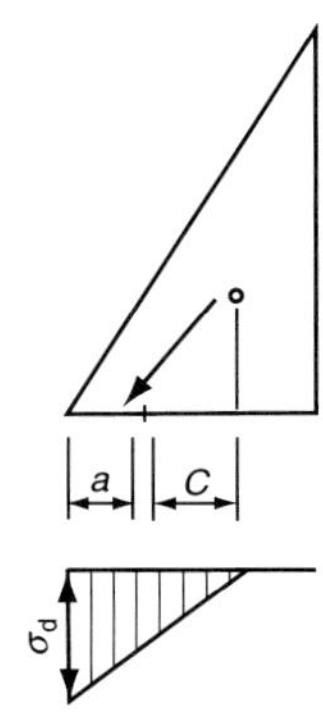

Fig. 118. Contact stresses at the dam base without tension in case of an earthquake directed downstream (reservoir full)

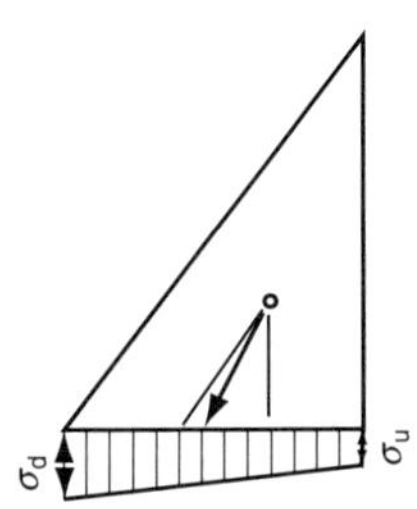

Fig. 119. Contact stresses at the dam base in case of an earthquake directed upstream (reservoir full)

- Downstream side:

$$\sigma_{\mathrm{d}}^{\mathrm{G-U+W-E}} = -\frac{215\cdot 0 - 21\cdot 8}{112}\left(1 + \frac{6(24\cdot 21 - 112/6)}{112}\right)$$

$$= -1\cdot 725\left(1 + \frac{5\cdot 54}{18\cdot 67}\right) = -2\cdot 24\ \mathrm{MN/m^2}$$

- Upstream side:

$$\sigma_{\mathrm{u}}^{\mathrm{G-U+W-E}} = -1\cdot 725(1 - 0\cdot 297) = -1\cdot 21\ \mathrm{MN/m^2}$$

6.2.2. Crest displacements

6.2.2.1. Hydrostatic pressure

- Deformation of concrete according to Equation (61):

$$w_{1\mathrm{c}} = \frac{\gamma_{\mathrm{w}} h^5}{E_{\mathrm{c}} B^3} + \frac{\gamma_{\mathrm{w}} h^3}{4 G_{\mathrm{c}} B} = \frac{0\cdot 01 \times 156^5}{20\,000 \times 112^3} + \frac{0\cdot 01 \times 156^3}{4 \times 8000 \times 112}$$

$$= 0\cdot 0329 + 0\cdot 0106 = 0\cdot 0435\ \mathrm{m}$$

- Length of dam's crest: $L = 960$ m
- Deformation of bedrock according to Equation (62):

$$w_{1\mathrm{r}} = \frac{3\gamma_{\mathrm{w}} h^4}{\pi E_{\mathrm{r}} B^2} + \frac{\gamma_{\mathrm{w}} h^2}{2 E_{\mathrm{r}}}\left(\frac{L}{B}\right)^{1/3}$$

$$= \frac{3 \times 0\cdot 01 \times 156^4}{\pi \times 15\,000 \times 112^2} + \frac{0\cdot 01 \times 156^2}{2 \times 15\,000}\left(\frac{960}{112}\right)^{1/3}$$

$$= 0\cdot 0301 + 0\cdot 0166 = 0\cdot 0467\ \mathrm{m}$$

- together:

$$w_1 = w_{1\mathrm{c}} + w_{1\mathrm{r}} = 0\cdot 0435 + 0\cdot 0467 = 0\cdot 0902\ \mathrm{m}$$

6.2.2.2. Rotation of valley floor due to reservoir load

- Settlement of valley floor according to Equation (64):

$$\max\ u_{\mathrm{r}} = 0\cdot 39\left(\frac{\gamma_{\mathrm{w}} h L}{E_{\mathrm{r}}}\right) = 0.39\left(\frac{0\cdot 01 \times 156 \times 960}{15\,000}\right) = 0\cdot 0389\ \mathrm{m}$$

- Negative (upstream) crest displacement (Fig. 61);

$$w_2 = \frac{\max\ u_r}{2} = -\frac{0{\cdot}0389}{2} = -0{\cdot}0195\ \text{m}$$

6.2.2.3. Temperature difference

- Difference of temperature between upstream and downstream faces of dam (measured at 1 m depth below concrete surface):

$$\Delta T_c = +1\ ^\circ\text{C}$$

- Angle of aperture of the triangular cross-section (Fig. 112) β:

$$\tan\beta = 0{\cdot}720 \quad \cos\beta = 0{\cdot}812 \quad \sin\beta = 0{\cdot}584$$

- Coefficient of thermal expansion: $\omega = 10^{-5}$ for concrete
- Crest displacement according to Equation (67):

$$w_3 = \frac{\omega \Delta T_c H}{2}\left(\frac{1}{\sin\beta\cos\beta} + \frac{1}{\tan\beta}\right)$$

$$= \frac{10^{-5} \times 1 \times 160}{2}\left(\frac{1}{0{\cdot}584 \times 0{\cdot}812} + \frac{1}{0{\cdot}720}\right) = 0{\cdot}0028\ \text{m}$$

In the case of a warmer downstream side, the crest displacement becomes negative.

6.2.2.4. Sum of crest displacements

- Hydrostatic pressure: $w_1 = 0{\cdot}0902$ m
- Reservoir load: $w_2 = -0{\cdot}0195$ m
- Total: $w_1 + w_2 = 0{\cdot}0707$ m
- Temperature difference: $w_3 = \pm 0{\cdot}0028$ m
- Extreme values: $\max w = 0{\cdot}0735$ m
  $\min w = -0{\cdot}0028$ m

6.2.3. Safety against sliding

6.2.3.1. Deadload, hydrostatic pressure, uplift and adhesion

- Vertical loads: $V = G - U = 215{\cdot}0 - 21{\cdot}8 = 193{\cdot}2$ MN/m
- Adhesion: $C = cB = 0{\cdot}1 \times 112 = 11{\cdot}2$ MN/m
- Horizontal loads: $H = W - C = 121{\cdot}7 - 11{\cdot}2 = 110{\cdot}5$ MN/m
- Base friction (artificial indentation of dam with bedrock) $\tan\phi = 1{\cdot}0$

- Factor of safety:

$$s = \frac{V}{H} \tan \phi = \frac{193 \cdot 2}{110 \cdot 5} \times 1 \cdot 0 = 1 \cdot 75 > 1 \cdot 5$$

6.2.3.2. Deadload, hydrostatic pressure, uplift, adhesion and earthquake

- Vertical loads: $V = G - U = 215{\cdot}0 - 21{\cdot}8 = 193{\cdot}2$ MN/m
- Horizontal loads:

$$H = W - C + E = 121 \cdot 7 - 11 \cdot 2 + 34 \cdot 0 = 144 \cdot 5 \text{ MN/m}$$

- Factor of safety:

$$s = \frac{193 \cdot 2}{144 \cdot 5} = 1 \cdot 34 > 1 \cdot 2$$

6.2.4. Conclusion

The gravity dam as designed is sufficiently safe.

6.3. Arch dams

The dam's geometry is defined by Figs 120 to 123. Due to the elasticity of the bedrock ($E_r = E_c = 20\ 000$ MN/m^2) the extreme contact stresses of the dam are about 50% higher than the calculated values based on the assumption of a linear stress distribution.

6.3.1. Deadload

Calculated tensile stresses due to a deadload of $\sigma^G < 1{\cdot}0$ MN/m^2 are tolerated in concrete dams.

6.3.2. Hydrostatic pressure

6.3.2.1. Crown cantilever

- Characteristic length according to Equation (110):

$$\lambda_x = 0 \cdot 76\sqrt{R_{yu} t_u} = 0 \cdot 76\sqrt{110 \times 42} = 51 \cdot 7 \text{ m}$$

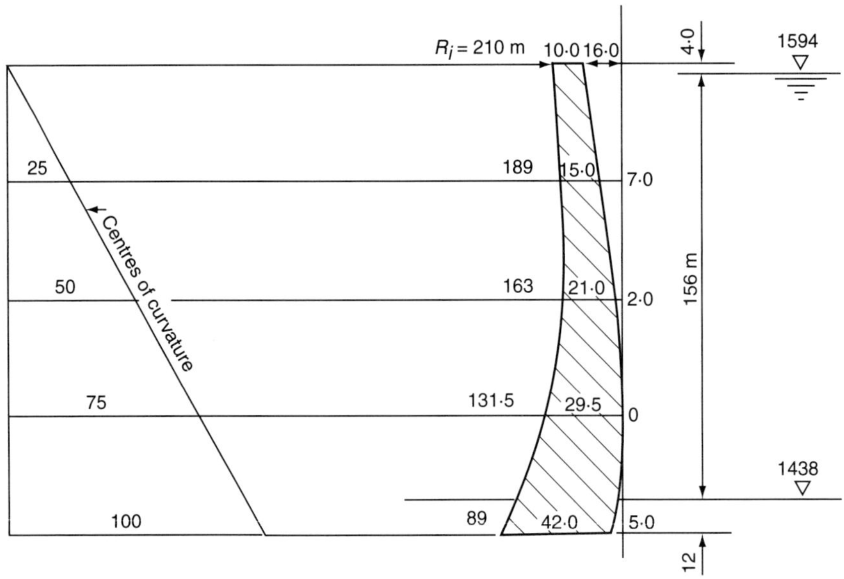

Fig. 120. Crown section of arch dam (cantilever)

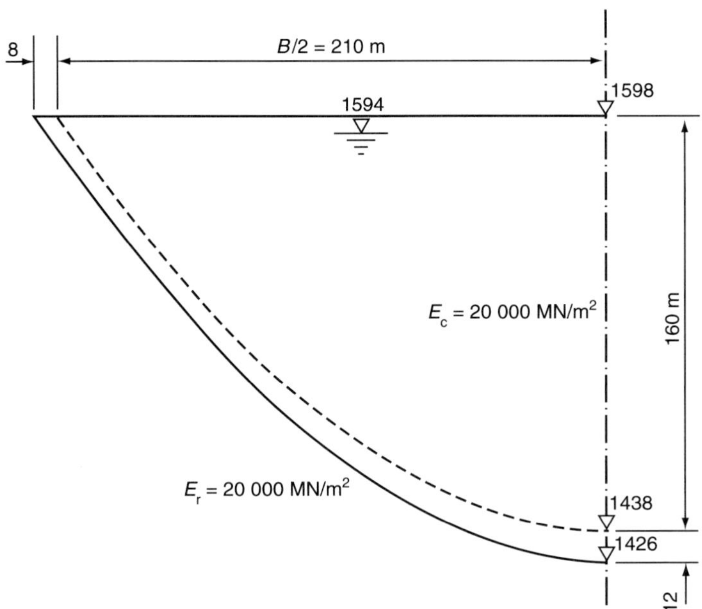

Fig. 121. Developed downstream view of arch dam

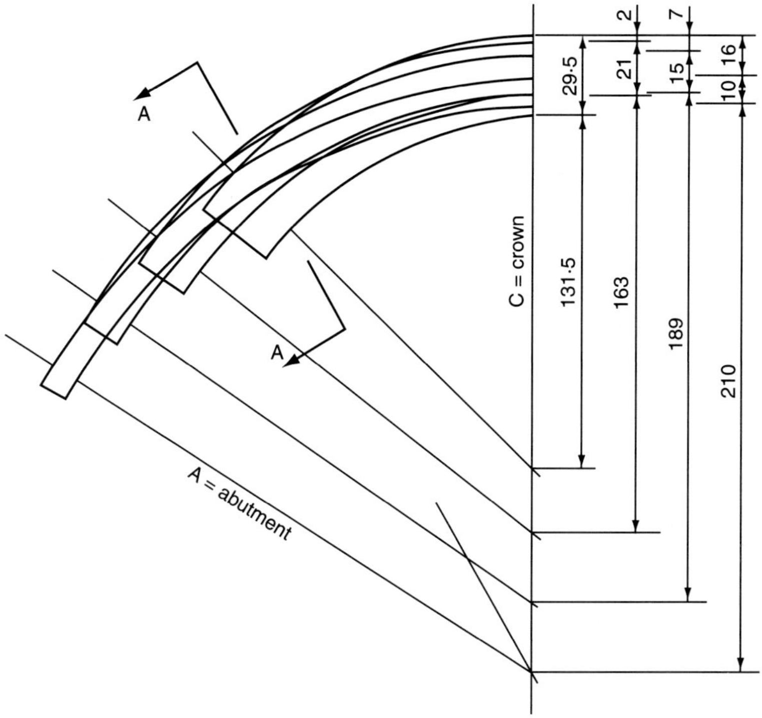

Fig. 122. Horizontal dam sections (arches)

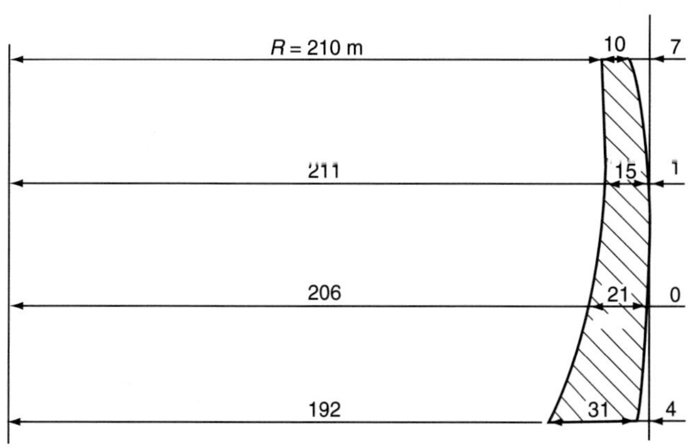

Fig. 123. Vertical section A–A of arch dam at the quarter-point of crest length (cantilever)

Table 29.

El.: m	t: m	g: kN/m^2	a_v: m	ga_v: M/Nm		ΔHga_v: MN m/m	
1598	10·0	240	21·0	5·04			0
					5·13	220·6	
1555	15·0	360	14·5	5·22			220·6
					5·76	247·7	
1512	21·0	504	12·5	6·30			468·3
					8·37	359·9	
1469	29·5	708	14·75	10·44			828·2
					18·33	788·2	
1426	42·0	1008	26·0	26·21			1616·4

$$\Delta = \frac{172}{4} = 43 \cdot 0 \text{ m} \quad \gamma_c = 24 \text{ kN/m}^3$$

$\Delta G = g\Delta H$: MN/m		a_s: m	$e = a_s - a_v$: m	$6e/t$	$\Delta G/t$: MN/m^2	σ^G: MN/m^2 down-stream	σ^G: MN/m^2 up-stream
	0	21·00	0	0	0	0	0
12·90							
	12·90	17·10	2·60	1·040	−0·860	−1·75	+0·03
18·58							
	31·48	14·88	2·38	0·680	−1·499	−2·52	−0·48
26·06							
	57·54	14·39	0·36	0·073	−1·951	−2·09	−1·81
36·89							
	94·43	17·12	−8·88	−1·269	−2·248	+0·60	−5·10

$$a_s = \frac{\sum \Delta Hga_v}{\Delta G}$$

- Fixing moment at the dam base according to Equation (111):

$$M_x^W = \frac{\gamma_w \lambda_x^2}{2}(H - \lambda_x) = -\frac{0 \cdot 01 \times 51 \cdot 7^2}{2}(168 - 51 \cdot 7)$$
$$= -1554 \text{ MNm/m}$$

- Shear force at the dam base according to Equation (112):

$$Q_x^W = \gamma_w \lambda_x \left(H - \frac{\lambda_x}{2}\right) = 0 \cdot 01 \times 51 \cdot 7\left(168 - \frac{51 \cdot 7}{2}\right)$$
$$= 73 \cdot 5 \text{ MN/m}$$

- Bending stresses:

$$\sigma_x^{\mathrm{W}} = \frac{6M_x^{\mathrm{W}}}{t_{\mathrm{u}}^2} = \mp\frac{6 \times 1554}{42{\cdot}0^2} = \mp 5{\cdot}29\ \mathrm{MN/m^2}$$

- Shear stress (parabolic distribution) according to Equation (125):

$$\tau_x^{\mathrm{W}} = \frac{3}{2}\left(\frac{Q_x^{\mathrm{W}}}{t_{\mathrm{u}}}\right) = \frac{3}{2}\left(\frac{73{\cdot}5}{42{\cdot}0}\right) = 2{\cdot}63\ \mathrm{MN/m^2}$$

- Normal stress due to deadload according to Table 29:

$$\sigma_x^{\mathrm{G}} = \frac{0{\cdot}60 - 5{\cdot}10}{2} = -2{\cdot}25\ \mathrm{MN/m^2}$$

- Principal stresses according to Equation (126):

$$\sigma_{1,2} = \frac{\sigma_x^{\mathrm{G}}}{2} \pm \left[\left(\frac{\sigma_x^{\mathrm{G}}}{2}\right)^2 + \tau_x^2\right]^{1/2} = -\frac{2{\cdot}25}{2} \pm \left[\left(\frac{2{\cdot}25}{2}\right)^2 + 2{\cdot}63^2\right]^{1/2}$$

$$= -1{\cdot}13 \pm 2{\cdot}86 = \begin{matrix} 1{\cdot}73 \\ -3{\cdot}99 \end{matrix}\ \mathrm{MN/m^2}$$

As the principal tensile stress is bigger than the tensile strength of the dam concrete C35 ($R_{\mathrm{c}} = 0{\cdot}6 \times 35 = 23\ \mathrm{MN/m^2}$)

$$R_{\mathrm{ct}} = \left(\frac{R_{\mathrm{c}}}{10}\right)^{1/2} = \left(\frac{23}{10}\right)^{1/2} = 1{\cdot}52\ \mathrm{MN/m^2}$$

cracks can develop in the interior of the dam.

6.3.2.2. *Arch at midheight*

- Characteristic length according to Equation (116):

$$\lambda_y = 0{\cdot}76\sqrt{R_{\mathrm{ym}} t_{\mathrm{m}}} = 0{\cdot}76\sqrt{173{\cdot}5 \times 21{\cdot}0} = 45{\cdot}9\ \mathrm{m}$$

- Average hydrostatic pressure according to Equation (114):

$$p_y = \gamma_{\mathrm{w}}\left(\frac{H}{2} - \lambda_x + \frac{\lambda_x^2}{2H}\right) = 0{\cdot}01\left(\frac{168}{2} - 51{\cdot}7 + \frac{51{\cdot}7^2}{2 \times 168}\right)$$

$$= 0{\cdot}403\ \mathrm{MN/m^2}$$

- The arch action in this dam according to Equation (115)

$$\eta = \frac{2p_y}{\gamma_w H} = \frac{2 \times 0{\cdot}403}{0{\cdot}01 \times 168} = 0{\cdot}48$$

is important.
- Horizontal fixing moment at the abutment according to Equation (117):

$$M_{yA}^{W} = p_y\left(\frac{\lambda_y^2}{2}\right) = -0{\cdot}403\left(\frac{45{\cdot}9^2}{2}\right) = -425 \text{ MN m/m}$$

- Normal force in horizontal direction according to Equation (118):

$$N_{ym}^{W} = p_y\left(R_{ym} + \frac{t_m}{2}\right) = -0{\cdot}403\left(173{\cdot}5 + \frac{21{\cdot}0}{2}\right)$$
$$= -74{\cdot}2 \text{ MN/m}$$

- Horizontal stresses at the abutment according to Equation (123):

$$\sigma_{yA}^{W} = \frac{N_{ym}^{W}}{t_m} \pm \frac{6M_{yA}^{W}}{t_m^2} = -\frac{74{\cdot}2}{21{\cdot}0} \pm \frac{6 \times 425}{21{\cdot}0^2}$$
$$= -3{\cdot}53 \pm 5{\cdot}78 = \begin{matrix} 2{\cdot}25 \\ -9{\cdot}31 \end{matrix} \text{ MN/m}^2 \begin{matrix} \text{upstream} \\ \text{downstream} \end{matrix}$$

In order to reduce the contact stresses in the abutment rock a downstream haunch is provided (Fig. 124).

- Eccentricity of pressure line: $e_0 = 425/74{\cdot}2 = 5{\cdot}73$ m
- Half of arch thickness at crown: $t_{mc}/2 = 21{\cdot}0/2 = 10{\cdot}50$ m
  Total $= 16{\cdot}23$ m
- Half of arch thickness at the abutment:
  $t_{mA}/2 = 26{\cdot}0/2 = -13{\cdot}00$ m
- Eccentricity at abutment: $e_A = 3{\cdot}23$ m
  is smaller than the kern width $t_{mA}/6 = 26{\cdot}0/6 = 4{\cdot}33$ m

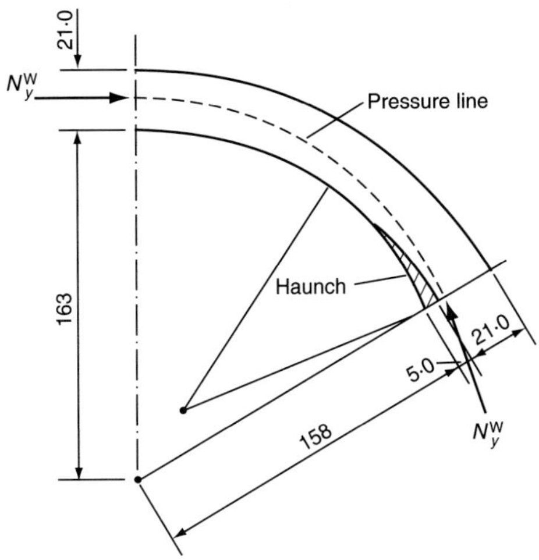

Fig. 124. Downstream haunch of the horizontal arch at midheight

The reduced abutment stresses (linear distribution) amount to

$$\sigma_{yA}^{W} = \frac{N_{ym}^{W}}{t_{mA}}\left(1 \pm \frac{6e_A}{t_{mA}}\right) = -\frac{74{\cdot}2}{26{\cdot}0}\left(1 \pm \frac{3{\cdot}23}{4{\cdot}33}\right)$$

$$= -2{\cdot}85(1 \pm 0{\cdot}746) = \begin{matrix} -0{\cdot}72 \\ -4{\cdot}98 \end{matrix}\ \text{MN/m}^2\ \begin{matrix} \text{upstream} \\ \text{downstream} \end{matrix}$$

With help of the normalized abscissa

$$\frac{y}{\lambda_y} = \frac{B_m}{2\lambda_y} = \frac{154{\cdot}15}{45{\cdot}9} = 3{\cdot}36$$

Fig. 80 gives the horizontal bending moment at the crown

$$M_{yC}^{W} = 0$$

and the normal stress amounts in this case to

$$\sigma_{yC}^{W} = \frac{N_{ym}^{W}}{t_{mC}} = -\frac{74{\cdot}2}{21{\cdot}0} = -3{\cdot}53\ \text{MN/m}^2$$

6.3.3. Variation of temperature

6.3.3.1. Crown cantilever

- Uniform variation of temperature according to Equation (1):

$$T_x = \frac{62}{t_{xu} + 2{\cdot}4} = \frac{62}{42 + 2{\cdot}4} = 1{\cdot}4\ ^\circ\text{C}$$

- Fixing moment at the dam base according to Equation (119):

$$M_x^{\text{T}} = \frac{E_c \omega T t_{xu}}{R_{yu}} \left(\frac{\lambda_x^2}{2}\right) = \frac{20\,000 \times 10^{-5} \times 1{\cdot}4 \times 42{\cdot}0}{110{\cdot}0} \left(\frac{51{\cdot}7^2}{2}\right)$$

$$= \pm 142{\cdot}9\ \text{MN/m}$$

- Bending stress:

$$\sigma_x^{\text{T}} = \frac{6M_x^{\text{T}}}{t_{xu}^2} = \frac{6 \times 142{\cdot}9}{42{\cdot}0^2} = \pm 0{\cdot}49\ \text{MN/m}^2$$

6.3.3.2. Horizontal arch at midheight

- $T_y = \dfrac{62}{21{\cdot}0 + 2{\cdot}4} = 2{\cdot}6\ ^\circ\text{C}$
- Fixing moment at the abutment according to Equation (120):

$$M_{yA}^{\text{T}} = \frac{E_c \omega T t_{m}}{R_{ym}} \left(\frac{\lambda_y^2}{2}\right) = \frac{20\,000 \times 10^{-5} \times 2{\cdot}6 \times 21{\cdot}0}{173{\cdot}5} \left(\frac{45{\cdot}9^2}{2}\right)$$

$$= \pm 66{\cdot}3\ \text{MN m/m}$$

- Normal force according to Equation (121):

$$N_{ym}^{\text{T}} = E_c \omega T t_{m} = 20\,000 \times 10^{-5} \times 2{\cdot}6 \times 21{\cdot}0 = \mp 10{\cdot}92\ \text{MN/m}$$

- Normal stresses at the abutment according to Equation (123):

$$\sigma_{yA}^{\text{T}} = \frac{N_{ym}^{\text{T}}}{t_{mA}} \pm \frac{6M_{yA}^{\text{T}}}{t_{mA}^2} = -\frac{10{\cdot}92}{26{\cdot}0} \pm \frac{6 \times 66{\cdot}3}{26{\cdot}0^2}$$

$$= -0{\cdot}42 \pm 0{\cdot}59 = \begin{matrix} 0{\cdot}17 \\ -1{\cdot}01 \end{matrix}\ \text{MN/m}^2 \begin{matrix} \text{downstream} \\ \text{upstream} \end{matrix}$$

- Normal stress at the crown:

$$\sigma_{yC}^{T} = \frac{N_{ym}^{T}}{t_{mC}} = -\frac{10\cdot 92}{21\cdot 0} = -0\cdot 52 \text{ MN/m}^2$$

6.3.4. *Shrinkage of concrete*

- $\varepsilon^S = \omega T/2$ approximately

6.3.5. *Earthquake*

- Horizontal response acceleration of the arch dam: $a_{Eh} = 0\cdot 1g$

6.3.5.1. *Reservoir empty*

- Crown cantilever:

$$\sigma_x^{E1} = \frac{a_{Eh}}{g}\left(\frac{\gamma_c}{\gamma_w}\right)\sigma_x^W = 0\cdot 1\left(\frac{0\cdot 024}{0\cdot 01}\right)5\cdot 29 = \pm 1\cdot 27 \text{ MN/m}^2$$

- Horizontal arch at midheight:

$$\sigma_{yA}^{E1} = \frac{a_{Eh}}{g}\left(\frac{\gamma_c}{\gamma_w}\right)\sigma_{yA}^W = 0\cdot 1\left(\frac{0\cdot 024}{0\cdot 01}\right)\begin{pmatrix}-0\cdot 72\\ -4\cdot 98\end{pmatrix} = \begin{matrix}-0\cdot 17\\ -1\cdot 20\end{matrix} \text{ MN/m}^2$$

$$\sigma_{yC}^{E1} = \frac{a_{Eh}}{g}\left(\frac{\gamma_c}{\gamma_w}\right)\sigma_{yC}^W = 0\cdot 24(-3\cdot 53) = -0\cdot 85 \text{ MN/m}^2$$

6.3.5.2. *Reservoir full*

- Crown cantilever:

$$\sigma_x^{E2} = \frac{a_{Eh}}{g}\left(\frac{\gamma_c + \gamma_w}{\gamma_w}\right)\sigma_x^W = 0\cdot 1\left(\frac{0\cdot 024 + 0\cdot 01}{0\cdot 01}\right)5\cdot 29$$
$$= \pm 1\cdot 80 \text{ MN/m}^2$$

- Horizontal arch at midheight:

$$\sigma_{yA}^{E2} = \frac{a_{Eh}}{g}\left(\frac{\gamma_c + \gamma_w}{\gamma_w}\right)\sigma_{yA}^{W} = 0{\cdot}1\left(\frac{0{\cdot}024 + 0{\cdot}01}{0{\cdot}01}\right)\begin{pmatrix}-0{\cdot}72\\ -4{\cdot}98\end{pmatrix}$$

$$= \begin{matrix}-0{\cdot}24\\ -1{\cdot}69\end{matrix}\ \text{MN/m}^2$$

$$\sigma_{yC}^{E2} = \frac{a_{Eh}}{g}\left(\frac{\gamma_c + \gamma_w}{\gamma_w}\right)\sigma_{yC}^{W} = 0{\cdot}34(-3{\cdot}53) = -1{\cdot}20\ \text{MN/m}^2$$

6.3.6. Stresses

6.3.6.1. Concrete stresses (linear distribution)

Table 30. Concrete stresses in MN/m^2

Loading		Cantilever		Abutment		
		Up-stream	Down-stream	Up-stream	Down-stream	Crown
Deadload		−5·10	+0·60	–	–	–
Shrinkage of concrete		+0·25	−0·25	−0·09	+0·51	+0·26
Variation of temperature		∓0·49	±0·49	±0·17	∓1·01	∓0·52
Hydrostatic pressure		+5·29	−5·29	−0·72	−4·98	−3·53
Normal Service	Summer	−5·34	+0·84	+0·08	−5·48	−3·79
	Winter	+0·93	−5·43	−0·98	+1·52	+0·78
Earth-quake	res. empty	±1·27	∓1·27	∓0·17	∓1·20	∓0·85
	res. full	±1·80	∓1·80	∓0·24	∓1·69	∓1·20
Normal service & earthquake	Summer	−6·61	+2·11	+0·25	+2·72	+1·63
	Winter	+2·73	−7·23	−1·22	−7·17	−4·98

6.3.6.2. Concrete strength

- Compressive strength of concrete C35:
  (a) static strength: $R_c^s = 0{\cdot}6 \times 35 = 23$ MN/m^2
  (b) dynamic strength: $R_c^d = 1{\cdot}25 \times 23 = 28{\cdot}7$ MN/m^2
- Tensile strength of concrete C35:
  (a) static strength: $R_{ct}^s = \sqrt{23/10} = 1{\cdot}52$ MN/m^2
  (b) dynamic strength: $R_{ct}^d = 1{\cdot}25 \times 1{\cdot}52 = 1{\cdot}90$ MN/m^2

6.3.6.3. *Commentary*

The static tensile strength of concrete is not surpassed in normal service, whereas the dynamic tensile strength is exceeded under earthquakes and causes repairable damage.

6.3.7. *Radial displacements*

6.3.7.1. *At midheight of dam*

- Due to hydrostatic pressure according to Equations (128) and (130):

$$w_{\mathrm{m}}^{\mathrm{W}} = \frac{N_{y\mathrm{m}}^{\mathrm{W}} R_{y\mathrm{m}}}{E_{\mathrm{c}} t_{\mathrm{mC}}} + \frac{9 M_{y\mathrm{A}}^{\mathrm{W}} \lambda_y^2}{E_{\mathrm{c}} t_{\mathrm{mA}}^3} = \frac{74{\cdot}2 \times 173{\cdot}5}{20\,000 \times 21{\cdot}0} + \frac{9 \times 425 \times 45{\cdot}9^2}{20\,000 \times 26{\cdot}0^3}$$
$$= 0{\cdot}0307 + 0{\cdot}0229 = 0{\cdot}0536 \text{ m}$$

- Due to temperature change according to Equations (129) and (130):

$$w_{\mathrm{m}}^{\mathrm{T}} = \omega T_y R_{y\mathrm{m}} + \frac{9 M_{y\mathrm{A}}^{\mathrm{T}} \lambda_y^2}{E_{\mathrm{c}} t_{\mathrm{mA}}^3}$$
$$= -10^{-5} \times 2{\cdot}6 \times 173{\cdot}5 - \frac{9 \times 66{\cdot}3 \times 45{\cdot}9^2}{20\,000 \times 26{\cdot}0^3}$$
$$= -0{\cdot}0045 - 0{\cdot}0036 = -0{\cdot}0081 \text{ m}$$

- Shrinkage of concrete:

$$\left(\varepsilon^{\mathrm{S}} = \frac{\omega T}{2}\right): \qquad w_{\mathrm{m}}^{\mathrm{S}} = \frac{w_{\mathrm{m}}^{\mathrm{T}}}{2} = -\frac{0{\cdot}0081}{2} = -0{\cdot}0041 \text{ m}$$

- Extreme values *without* earthquake:

$$\max\ w_{\mathrm{m}} = w_{\mathrm{m}}^{\mathrm{W}} - w_{\mathrm{m}}^{\mathrm{T}} + w_{\mathrm{m}}^{\mathrm{S}} = 0{\cdot}0536 + 0{\cdot}0081 - 0{\cdot}0041$$
$$= 0{\cdot}0576 \text{ m}$$

$$\min\ w_{\mathrm{m}} = w_{\mathrm{m}}^{\mathrm{T}} + w_{\mathrm{m}}^{\mathrm{S}} = -0{\cdot}0081 - 0{\cdot}0041 = -0{\cdot}0122 \text{ m}$$

- Earthquake:

(a) reservoir empty: $w_{\mathrm{m}}^{\mathrm{E1}} = \dfrac{a_{\mathrm{Eh}}}{g}\left(\dfrac{\gamma_{\mathrm{c}}}{\gamma_{\mathrm{w}}}\right) w_{\mathrm{m}}^{\mathrm{W}}$

$$= \pm 0{\cdot}1\left(\frac{0{\cdot}024}{0{\cdot}01}\right)0{\cdot}0536 = \pm 0{\cdot}0129 \text{ m}$$

(b) reservoir full: $w_{\mathrm{m}}^{\mathrm{E2}} = \dfrac{a_{\mathrm{Eh}}}{g}\left(\dfrac{\gamma_{\mathrm{c}} + \gamma_{\mathrm{w}}}{\gamma_{\mathrm{w}}}\right) w_{\mathrm{m}}^{\mathrm{W}}$

$$= \pm 0{\cdot}1\left(\frac{0{\cdot}024 + 0{\cdot}01}{0{\cdot}01}\right)0{\cdot}0536$$

$$= \pm 0{\cdot}0182 \text{ m}$$

- Extreme values *with* earthquake:

$$\max\ w_{\mathrm{m}} = w_{\mathrm{m}}^{\mathrm{W}} - w_{\mathrm{m}}^{\mathrm{T}} + w_{\mathrm{m}}^{\mathrm{S}} + w_{\mathrm{m}}^{\mathrm{E2}}$$
$$= 0{\cdot}0536 + 0{\cdot}0081 - 0{\cdot}0041 + 0{\cdot}0182 = 0{\cdot}0758 \text{ m}$$

$$\min w_{\mathrm{m}} = w_{\mathrm{m}}^{\mathrm{T}} + w_{\mathrm{m}}^{\mathrm{S}} - w_{\mathrm{m}}^{\mathrm{E1}}$$
$$= -0{\cdot}0081 - 0{\cdot}0041 - 0{\cdot}0129 = -0{\cdot}0251 \text{ m}$$

6.3.7.2. At dam base

- Developed length of arch dam foundation: $L_{\mathrm{r}} = 578$ m
- Owing to hydrostatic pressure according to Equation (59):

$$w_x^{\mathrm{W}} = \frac{Q_x^{\mathrm{W}}}{E_{\mathrm{r}}}\left(\frac{L_{\mathrm{r}}}{t_x}\right)^{1/3} = \frac{73{\cdot}5}{20\,000}\left(\frac{578}{42{\cdot}0}\right)^{1/3} = 0{\cdot}0088 \text{ m}$$

6.3.8. Safety of rupture

Based on the assumption of a sufficient strength of bedrock.

6.3.8.1. Arch at midheight of dam

- Average horizontal stress:

$$\sigma_{y\mathrm{m}}^{\mathrm{W}} = \frac{\gamma_{\mathrm{w}} h}{2}\left(\frac{R_{y\mathrm{m}} + t_{\mathrm{mC}}/2}{t_{\mathrm{mC}}}\right) = -\frac{0{\cdot}01 \times 168}{2}\left(\frac{173{\cdot}5 + 21{\cdot}0/2}{21{\cdot}0}\right)$$
$$= -7{\cdot}36 \text{ MN/m}^2$$

- Safety of concrete:

$$s = \frac{R_s}{\sigma_{ym}^{W}} = \frac{23}{7{\cdot}36} = 3{\cdot}13 > 2{\cdot}0$$

6.3.8.2. Bedrock at abutment (Fig. 125)

- $\cos\left(\frac{\pi}{2} - \beta\right) = \cos\,(90° - 68°) = \cos\,22° = 0{\cdot}9272$
- $\cos\,\alpha = \cos\,16° = 0{\cdot}9613$
- Contact stresses (normal to the abutment):

$$\sigma_{yA}^{W} = \frac{N_{ym}^{W}}{t_{mA}} \times \cos\left(\frac{\pi}{2} - \beta\right) \times \cos\,\alpha\left(1 \pm \frac{6c}{t_{mA}} \times \cos\alpha\right)$$

$$= -\frac{74{\cdot}2}{26{\cdot}0} \times 0{\cdot}9272 \times 0{\cdot}9613\left(1 \pm \frac{6 \times 3{\cdot}13}{26{\cdot}0}\right)$$

$$= -2{\cdot}54(1 \pm 0{\cdot}722) = \begin{matrix} -4{\cdot}37 \\ -0{\cdot}71 \end{matrix}\ \text{MN/m}^2\ \begin{matrix} \text{downstream} \\ \text{upstream} \end{matrix}$$

- Coefficient of friction for concrete indentated with bedrock: $\tan\,\phi = 1{\cdot}0$
- Safety against sliding of arch at abutment:

$$s = \frac{\cos(\frac{\pi}{2} - \beta)}{\sin(\frac{\pi}{2} - \beta)} \times \tan\,\phi = \frac{0{\cdot}9272}{0{\cdot}3746} \times 1{\cdot}0 = 2{\cdot}48 > 1{\cdot}5$$

6.3.9. Conclusion

The arch dam as designed is sufficiently safe.

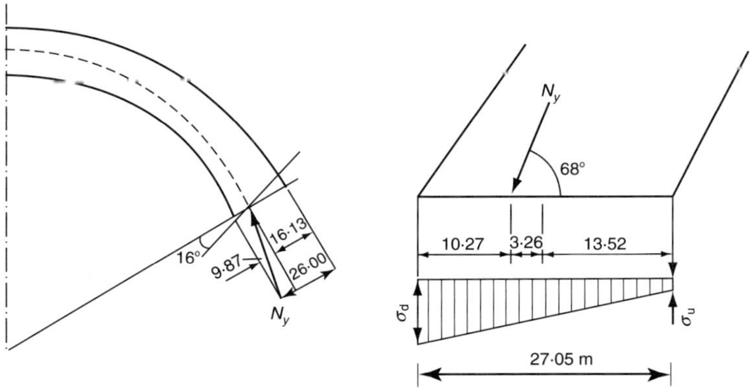

Fig. 125. Contact stresses at the abutment in midheight of arch dam

Appendix: Unit conversion table

Quantity	American units	SI units
Length	1 inch	= 25·4 millimetres (mm)
		= 2·54 centimetres (cm)
	1 foot	= 0·3048 metres (m)
	1 yard	= 0·9144 m
	1 mile	= 1609 m
		= 1·609 km
Area	1 square inch	= 645·2 mm^2 = 6·452 cm^2
	1 square foot	= 929 cm^2 = 0·0929 m^2
	1 square yard	= 0·8361 m^2
	1 acre	= 4047 m^2
	1 square mile	= 2·590 km^2
Volume or section modulus	1 cubic inch	= 16 390 mm^3 = 16·39 cm^3
	1 cubic foot	= 0·02832 m^3 = 28·32 litres (l)
	1 cubic yard	= 0·7646 m^3 = 764·6 l
	1 US gallon	= 3·785 l
Second moment of inertia	1 inch^4	= 416 200 mm^4 = 41·62 cm^4
	1 foot^4	= 0·008631 m^4
	1 yard^4	= 0·6991 m^4
Mass	1 ounce	= 28·35 grams (g)
	1 pound	= 453·6 g = 0·4536 kilograms (kg)
	1 US short ton	= 907·2 kg
Force or weight	1 pound (lb)	= 4·445 newton (N)
	1 kilopound (kip)	= 4·445 kilonewton (kN)
	1 US short ton	= 8·890 kN
	1000 US short tons	= 8·890 meganewtons (MN)
Pressure or stress	1 lb/in^2 (psi)	= 0·006 890 N/mm^2
		= 0·6890 N/cm^2
		= 6890 N/m^2 = 6·890 kN/m^2
	1000 lbs/in^2 (ksi)	= 6·890 N/mm^2 = 6·890 MN/m^2
		= 689·0 N/cm^2 = 0·6890 kN/cm^2
	1 lb/ft^2 (psf)	= 47·85 N/m^2 = 0·047 85 kN/m^2

Quantity	American units	SI units
Density	1 lb/in^3	$=0.00027125$ N/mm^3
		$=0.27125$ N/cm^3
		$=271\ 250$ N/m^3 $=271.250$ kN/m^3
	1000 lbs/ft^3	$=157.0$ kN/m^3
	1 US short ton/ft^3	$=314.0$ kN/m^3
	1 US short ton/yd^3	$=11.63$ kN/m^3
Velocity	1 in/second	$=25.2$ mm/s
	1 ft/minute	$=5.083$ mm/s
	1 mile/hour (mph)	$=1.609$ km/h $=0.4470$ m/s
Rate of flow	1 US gallon/hour	$=3.785$ l/h $=1.051$ ml/s
	1 US gallon/minute	$=3.785$ l/min $=63.08$ ml/s
	1 ft^3/second	$=0.02832$ m^3/s $=101.94$ m^3/h
	1 yd^3/second	$=0.7646$ m^3/s
	1 acrefoot/second	$=1233.5$ m^3/s
Power	1 horsepower (hp)	$=745.7$ watt (W)
		$=0.7457$ kilowatt (kW)
Earth's gravitation	$1g=32.15$ ft/s^2	$=9.8$ m/s^2
Temperature	F in °F	$=\frac{5}{9}(F-32)$ °C

References

1. Garbrecht, G. (ed.) (1987). Historische Talsperren. *Deutscher Verband für Wasserwirtschaft und Kulturbau.* Wittwer, Stuttgart.
2. Herzog, M. (1993). Historische Entwicklung der Talsperrenberechnung. *Bautechnik* **70**, 576 – 586.
3. Fellenius, W. (1927). *Erdstatische Berechnungen mit Reibung und Kohäsion unter Annahme kreiszylindrischer Gleitflächen.* Ernst & Sohn, Berlin.
4. Collin, A. (1846). *Recherches Expérimentales sur les Glissements Spontanés de Terrains Argileux, Accompagnées de Considérations sur Quelques Principes de Mécanique Terrestre.* Carilian-Goeury, Paris.
5. Cooling, L. F. and Golder, H. Q. (1942). The analysis of the failure of an earth dam during construction. *Journal ICE* **19**, 36 – 55.
6. Londe, P. (1965). Une méthode d'analyse à trois dimensions de la stabilité d'une rive rocheuse. *Annales Ponts et Chaussées* **135**, 37 – 60.
7. Kovari, K. and Fritz, P. (1976). Stabilitätsberechnung ebener und räumlicher Felsböschungen. *Rock Mechanics* **8**, 73 – 113.
8. Herzog, M. (1981). Die Standsicherheit von Erd- und Felsböschungen bei beliebiger Form der Rutschfläche. *Bauingenieur* **56**, 89 – 95.
9. Herzog, M. (1980). Die Setzungen des 244 m hohen Mica-Damms in Kanada unter Eigengewicht nach Messung und Rechnung. *Bauingenieur* **55**, 427 – 431.
10. Schober, W. (1967). Behaviour of Gepatsch rockfill dam. *9th ICOLD, Istanbul.* Vol. III, pp. 677 – 699 (Q 34/R 39).
11. Schwab, H. H. (1979). Staudamm Gepatsch — Analyse zum langjährigen Verhalten. *Österr. Wasserwirtschaft* **31**, 202 – 210.
12. Herzog, M. (1988). Risse in Staudämmen, warum? *Wasserwirtschaft* **78**, 300 – 305.
13. Schober, W. (1970). The interior stress distribution of the Gepatsch rockfill dam. *10th ICOLD, Montreal.* Vol. I, pp. 169 – 187 (Q 36/R 10).
14. Herzog, M. (1992). Der Bruch des Sheffield-Damms in Santa Barbara beim Erdbeben am 29. Juni 1925 im Rückblick. *Österr. Wasserwirtschaft* **44**, 206 – 208.
15. Clough, R. W. and Woodward, R. J. (1967). Analysis of embankment stresses and deformations. *Proc. ASCE* **93**, SM 4, pp. 529 – 541.
16. Lefebvre, G., Duncan, J. M. and Wilson, E. L. (1973). Three-dimensional finite element analysis of dams. *Proc. ASCE* **99**, SM 7, 495 – 507.
17. Eisenstein, Z. and Simmons, J. V. (1975). Three-dimensional analysis of Mica Dam. *Proceedings of the International Conference on Criteria and Assumptions in the Numerical Analysis of Dams, Swansea.* pp. 1052 – 1069.

18. Finn, W. D. L. (1990). Seismic analysis of embankment dams. *Dam Engineering* **1**, No. 1, 59 – 75.
19. Schober, W., Hammer, H. and Hupfauf, B. (1989). Load transfer in embankment dams — model testing. *12th ICSMFE, Rio de Janeiro*. Paper 11/24.
20. Schober, W. and Hupfauf, B. (1991). The bearing behaviour of embankment dams in narrow valleys. *17th ICOLD, Vienna*. (Q 67/R 5).
21. Weber-Ebenhof, A. (1895). Der Bruch der Staumauer von Bouzey und die daraus für die Jaispitzbach-Reservoire zu ziehenden Lehren. *Österr. Monatsschrift f.d. öffentl. Baudienst* **1**, 161 – 167.
22. Rankine, W. J. M. (1872). Report on the design and construction of masonry dams. *The Engineer* **33**, 1 – 2.
23. Lévy, M. (1898). Sur l'équilibre élastique d'un barrage en maçonnerie à section triangulaire. *Comptes Rendues* **127**, 10 – 24.
24. Wilson, J. S. and Gore, W. (1905). Stresses in dams — an experimental method of investigating the distribution of stresses in dams. *Engineering*, **80**, 105 – 107.
25. Wilson, J. S. and Gore, W. (1907 – 1908). Stresses in dams — an experimental investigation by means of India rubber models. *Minutes of Proc. ICE*, Part. II, **172**, pp. 134 – 135.
26. Richardson, L. F. (1909 – 1911). The approximate arithmetical solution by finite differences of physical problems involving differential equations, with an application to stresses in masonry dams. *Philosophical Transactions* A **210**, 307 – 320.
27. Kiel, J. (1889). Zur Berechnung von Wasserdruckmauern, insbesondere von Talsperren. *Centralblatt d. Bauverwaltung* **9**, 397 – 398.
28. Lieckfeldt, H. (1898). Die Standfestigkeit von Staumauern mit offenen Lagerfugen. *Centralblatt d. Bauverwaltung* **18**, 105 – 111.
29. Fillunger, P. (1913). Der Auftrieb in Talsperren. *Österr. Wochenschrift f.d. öffentl. Baudienst* **65**, 532 – 556, 567 – 570.
30. Terzaghi, K. (1934). Beanspruchung von Gewichtsstaumauern durch das strömende Sickerwasser. *Bautechnik* **12**, 379 – 382.
31. Heinrich, G. (1938). Wissenschaftliche Grundlagen der Theorie der Setzung von Tonschichten. *Wasserkraft & Wasserwirtschaft* **33**, 5 – 10.
32. Brahtz, J. H. A. (1936). The stress function and photoelasticity applied to dams. *Proc. ASCE* **71**, Paper 1949.
33. Tölke, F. (1938). *Talsperren*. Springer, Berlin.
34. Zienkiewicz, O. C. (1947). The stress distribution in gravity dams. *Journal ICE*, 244 – 250 (Paper 5531).
35. Southwell, R. V. (1940). *Relaxation Methods in Engineering Science*. Clarendon Press, Oxford.
36. Stucky, A. (1954). Quelques problèmes relatifs aux fondations des grands barrages-réservoirs — Barrages de Mauvoisin et de la Grande Dixence. *Bulletin Technique de la Suisse Romande* **80**, 317 – 325, 329 – 336.
37. Pearce, C. E. (1940). Design of Hiwassee Dam — basic considerations. *Civil Engineering ASCE* **10**, 340 – 342.
38. Rescher, O. J. (1965). *Talsperrenstatik — Berechnung und Bemessung von Gewichtsstaumauern*. Springer, Berlin.
39. Turner, M. J., Clough, R. W., Martin, H. C. and Topp, L. J. (1956). Stiffness and deflection analysis of complex structures. *Journal of Aero Sciences* **23**, 805 – 823.

40. Clough, R. W. (1960). The finite element method in plane stress analysis. *Proceedings 2nd ASCE Conference on Electronic Computation.*
41. Campbell, J. S. and Zienkiewicz, O. C. (1975). Gravity dam analysis — the effect of valley width. *Proceedings of the International Conference on Criteria and Assumptions in the Numerical Analysis of Dams, Swansea* pp. 1007 – 1031.
42. Herzog, M. (1989). Horizontale Tragwirkung gerader Gewichtsstaumauern in engen Tälern. *Schweizer Ingenieur & Architekt* **107**, 111 – 115, 341.
43. Ritter, H. (1913). *Die Berechnung von bogenförmigen Staumauern.* Lang, Stuttgart.
44. Vischer, H. and Wagoner, L. (1889). On strains in curved masonry dams. *Transactions of the Technical Society of the Pacific Coast* **6**, 75 – 115.
45. Wisner, G. Y. and Wheeler, T. E. (1905). Investigation of stresses in high masonry dams of short span. *Engineering News* **54**, 141 – 144.
46. Jorgensen, L. R. (1914). The constant angle arch dam. *Transactions of the ASCE* **67**, Paper 1322, 685 – 721.
47. Eastwood, J. S. (1910). An arch dam design for the site of the Shoshone Dam. *Engineering News* **63**, 678 – 680.
48. Stucky, A. (1922). Etude sur les barrages arqués. *Bulletin Technique de la Suisse Romande* **48**, 1 – 7, 25 – 30, 49 – 53, 85 – 90, 97 – 103.
49. Howell, C. H. and Jaquith, A. C. (1929). Analysis of arch dams by the trial load method. *Transactions of the ASCE* **93**, Paper 1712, 1191 – 1316.
50. Vogt, F. (1925). Über die Berechnung der Fundamentdeformation. *Avhandlinger, Norske Videnskaps. Akademi, I. Math.-Naturv. Klasse No. 2.* Dybvad, Oslo.
51. Boussinesq, J. (1885). *Application des Potentiels à l'étude de l'équilibre et du Mouvement des Solides Élastiques.* Gauthier-Villars, Paris.
52. Noetzli, F. A. (1926). Über die Versuchs-Gewölbe-Staumauer am Stevenson Creek in Californien. *Schweiz. Bauzeitung* **87**, 13 – 16.
53. Houk, I. E. (1932). Technical design studies for Hoover Dam. *Western Construction News* 10 April 1932, 187 – 193.
54. Savage, J. L. and Houk, I. E. (1932). Model tests confirm design of Hoover Dam. *Engineering News Record* **108**, 494 – 499.
55. Jurecka, W. (1948). Die Berechnung bogenförmiger Staumauern nach dem Lastaufteilungsverfahren. *Österr. Bauzeitschrift* **4**, 173 – 175, 195 – 203.
56. Beaujoint, N. (1965). New developments in traditional methods of analysis of arch dams. *International Symposium on Theory of Arch Dams, Southhampton* Pergamon, Oxford, pp. 415 – 440.
57. Comité français des grands barrages (1979): Désordres graves constatés sur des barrages français. *13th ICOLD, New Delhi* Vol. II, pp. 557 – 584 (Q 49/R 37).
58. Smith, B. A. (1919 – 1920). Arched dams. *Transactions of the ASCE* **83**, Paper 1461, 2027 – 2093.
59. Tschech, E. and Jaburek, F. (1951). Berechnungen von Bogenstaumauern im Vergleich mit den Ergebnissen statischer Modellversuche. *Österr. Bauzeitschrift* **6**, 17 – 23.
60. Rescher, O. J. (1951). Die Gewölbestaumauer — Berechnung einer Kreiszylinderschale mittels Relaxationsmethode. *Abhandlungen d. Dokumentationszentrums d. Technik* No. 1. Vienna.
61. Lombardi, J. (1955). *Les Barrages en Voûte Mince.* Rouge, Lausanne.
62. Allen, D. N. de G., Chitty, L., Pippard, A. J. S. and Severn, R. T. (1956). The experimental and mathematical analysis of arch dams — with special reference to Dokan. *Proceedings of the ICE* Part 1, **5**, Paper 6113, 198 – 258.

63. Gilg, B. (1970). *Die Staumauer Punt dal Gall. Schweiz. Talsperrentechnik* Schweiz. Wasserwirtschaftsverband, Baden, pp. 75 – 82.
64. Herzog, M. (1992). Rasche Vorbemessung von Bogenstaumauern nach der Schalentheorie. *Österr. Ing. & Arch.-Zeitschrift* **137**, 162 – 171; Herzog, M. (1995). *Österr. Ing. & Arch.-Zeitschrift* **140**, 117.
65. Taig, I. C. (1962). *Structural Analysis by the Matrix Displacement Method.* English Electric Aviation Ltd, Stress Office Report No. 17.
66. Ergatoudis, J., Irons, B. M. and Zienkiewicz, O. C. (1968). Three-dimensional analysis of arch dams and their foundations. *Symposium on Arch Dams, London.* Institution of Civil Engineers, London, Paper 4, pp. 37 – 50.
67. Ergatoudis, J., Irons, B. M. and Zienkiewicz, O. C. (1962). Curved isoparametric quadrilateral elements for finite element analysis. *International Journal of Solids & Structures* **4**, 31 – 39.
68. Wittke, W. and Gell, K. (1984). Wechselwirkung zwischen Staumauer und Untergrund. *Wasserwirtschaft* **74**, 137 – 141.
69. Gell, K. and Wittke, W. (1986). A new design concept for arch dams taking into account seepage forces. *Rock Mechanics & Rock Engineering* **19**, 187 – 204.
70. O'Connor, J. P. F. (1985). The modelling of cracks, potential crack surfaces and construction joints in arch dams by curved surface interface elements. *15th ICOLD, Lausanne* Vol. II, pp. 389 – 406 (Q 57/R 22).
71. Clough, R. W., Raphael, J. M. and Mojtahedi, S. (1971). *ADAP — A Computer Program for Static and Dynamic Analysis of Arch Dams.* University of California, Berkeley, USA, Report EERC 73-14.
72. Stevenson, R., Bowes, D. E. and Radochia, J. P. (1986). Finite element analysis of arch dams on a personal computer. *Water Power & Dam Construction* **38**, No. 12, 19 – 22.
73. Herzog, M. (1993). Eine Checkliste zur Beurteilung der Sicherheit von alten und neuen Talsperren. *Bauingenieur* **68**, 477 – 481.
74. Noetzli, F. A. (1922). Arch dam temperature changes and deflection measurements. *Engineering News Record* **89**, 930 – 932.
75. Houk, I. E. (1930). Temperature variation in concrete dams. *Western Construction News* 601 – 608.
76. Serafim, J. L. (1968). Discussion of Session 3. *Symposium on Arch Dams, London.* Institution of Civil Engineers, London, pp. 152 – 153.
77. Marcello, C. (1956). Isolato double-curvature arch dam. *Proceedings of the ASCE* **82**, PO 3, Paper 995, pp. 1 – 19.
78. Gilg, B. (1967). Measures à la longue échéance au barrage voûte de Mauvoisin. *9th ICOLD, Istanbul* Vol. III, pp. 629 – 643 (Q 34/R 35).
79. Herzog, M. (1980). Failure mechanisms of dams subjected to earthquakes. *Water Power & Dam Construction* **32** No. 8, 28 – 30.
80. Saini, S. S. *et al.* (1972). Behavior of Koyna Dam — Dec. 11, 1967 earthquake. *Proceedings of the ASCE* **98**, ST 7, 1395 – 1412.
81. Swanson, A. A. & Sharma, R. P. (1979). Effects of the 1971 San Fernando earthquake on Pacoima arch dam. *13th ICOLD, New Delhi*, Vol. III, pp. 797 – 824 (Q 51/R 3).
82. Housner, G. W. and Jennings, P. C. (1972). The San Fernando California earthquake. *Earthquake Engineering & Structural Dynamics* **1**, No. 1, 5 – 31.
83. Herzog, M. (1984). Die einfachste Näherungsberechnung von Dammbruchwellen. *Wasserwirtschaft* **74**, 583 – 585.
84. Herzog, M. (1989). The influence of reservoir loading on dam stresses and deformations. *Water Power & Dam Construction* **41**, No. 7, 26 – 28.

85. Ludescher, H. (1990). Die Sanierung der Kölnbreinsperre. *Österr. Ing. & Arch.-Zeitschrift* **135**, 17 – 25; Ludescher, H. (1994). *Österr. Ing. & Arch.-Zeitschrift* **139**, 202 – 212.
86. Herzog, M. (1988). Die Baugrundverformung bei Gewölbestaumauern. *Bautechnik* **65**, 376 – 380.
87. Kettner, R. (1967). Die Fundamentkräfte der Gewölbesperren. *Felsmechanik & Ingenieurgeologie* Suppl. III, 66 – 73.
88. Herzog, M. (1990). Die statische Wirkung des Dichtungsschleiers bei Talsperren. *Wasserwirtschaft* **80**, 29 – 31.
89. Rescher, O. J. (1981). Geomechanische Modellversuche für die Gründung von Talsperren. *Rock Mechanics* **14**, 117 – 166.
90. *ICOLD Congresses*. Held every three years in a different country, e.g., 1985 Lausanne, 1988 San Francisco, 1991 Vienna and 1994 Durban.
91. Ziegler, P. (1927). *Der Talsperrenbau*, Ernst & Sohn, Berlin, Vol. II.
92. Anonymous (1989). Bruch der Talsperre oberhalb Johnstown in Pennsylvanien. *Centralblatt d. Bauverwaltung* **9**, 250 – 252, 266 – 267.
93. Stucky, A. (1924). Der Talsperrenbruch im Val Gleno. *Schweiz. Bauzeitung* **83**, 63 – 67, 74 – 76, 295 – 296.
94. Wiley, A. J. (1928). *Causes Leading to the Failure of St. Francis Dam.* California State Printing Office, Sacramento, USA.
95. Middlebrooks, T. A. (1942). Fort Peck slide. *Transactions of the ASCE* **107**, 723 – 764.
96. Bellier, J. (1967). Le barrage de Malpasset. *Travaux* No. 385, 363 – 383.
97. Ru Naihua (1989). Analysis of an experimental arch dam failure. *Water Power & Dam Construction* **41**, No. 7, 30 – 33.
98. Anonymous (1964). *Investigation of Failure, Baldwin Hill Reservoir, Los Angeles.* California Department of Water Resources, Sacramento, USA.
99. Seed, H. B. *et al.* (1975). The slides in the San Fernando dams during the earthquake of February 9, 1971. *Proceedings of the ASCE* **101**, GT 7, 651 – 688.
100. Chadwick, W. L. (1976). *Report of the Panel to Review Causes of Teton Dam Failure in 1976.* Government Printing Office, Washington, USA.
101. Comité National Suisse des Grands Barrages (1985). *Barrages Suisses – Surveillance et Entretien.* Zurich, pp. 189 – 192.
102. Herzog, M. (1995). Statische Lehren aus dem Fall Kölnbrein. *Bautechnik* **72**, 441 – 445.
103. Herzog, M. (1993). Die Bau- und Konsolidationssetzungen von Staudämmen. *Wasser Energie Luft* **85**, 35 – 38.
104. Taylor, D. W. (1947). *Review of Pressure Distribution Theories, Earth Pressure Cell Investigations and Pressure Distribution Data.* Waterways Experimental Station, Vicksburg, Soil Mechanics Fact Finding survey — Progress Report.
105. Janssen, H. A. (1895). Versuche über den Getreidedruck in Silozellen. *VDI-Zeitschrift* **39**, 1045 – 1049.
106. Webster, J. L. (1970). Mica Dam designed with special attention to control of cracking. *10th ICOLD, Montreal* vol. I, pp. 73 – 93 (Q 36/R 5).
107. Lauffer, H. and Schober, W. (1964). The Gepatsch rockfill dam in the Kauner Valley. *8th ICOLD, Edinburgh* Vol. II, pp. 635 – 660 (Q 31/R 4).
108. Eisenstein, Z. and Law, S. T. C (1977). Analysis of consolidation behavior of Mica Dam. *Proceedings of the ASCE* **103**, GT 8, 879 – 895.
109. Schwinn, K. (1970). Beitrag zur Berechnung erdbebensicherer Staudämme. *Strassen- & Tiefbau* **5**, 396 – 403.

110. Herzog, M. (1985). Closed formulae for earthquake induced deformations in dams. *Water Power & Dam Construction* **37**, No. 1, 33 – 34.
111. Herzog, M. (1980). Die Beanspruchung von Staudämmen durch Erdbeben. *Bauingenieur* **55**, 51 – 55.
112. Herzog, M. (1992). Zutreffende Erdbebenbemessung von Talsperren mit einem verbesserten statischen Ersatzlastverfahren. *Österr. Wasserwirtschaft* **44**, 15 – 21.
113. Herzog, M. (1994). Dreidimensionale Trägerrostberechnung von Staudämmen in engen Tälern unter Erdbebenbeanspruchung. *Österr. Wasserwirtschaft* **46**, 224 – 229.
114. Austrian National Committee of Large Dams (1991). Dams in Austria. *Die Talsperren Österreichs* No. 32. Österr. Wasserwirtschaftsverband, Vienna.
115. Petzny, H., Schober, W. and Widmann, R. (1977). Messungen an österreichischen Talsperren. *Die Talsperren Österreichs* No. 23. Österr. Wasserwirtschaftsverband, Vienna.
116. Anonymous (1962). Oroville's low bid: $121 million. *Engineering News Record* **169**, No. 5, 19 – 21.
117. Schnitter, E. (1962). Der Erddamm Oroville in Kalifornien. *Schweiz. Bauzeitung* **80**, 779 – 780.
118. Mackintosh, I. B. (1964). Mammoth operations for earth-moving on Oroville contract. *Civil Engineering* **60**, 1229 – 1233.
119. Toth, I. (1972). Le barrage Oroville en Californie (Etats-Unis). *La Technique des Travaux* No. 336, pp. 241 – 260.
120. Maidl, B. (1980). Sicherheitsfragen verhindern den Weiterbau der längsten doppelt gekrümmten Bogenstaumauer der Welt am Auburn-Folsom-South Projekt in Kalifornien. *Bauingenieur* **55**, 155 – 156.
121. Herzog, M. (1992). Eigenschwingungsdauer sowie dynamischer Gleitmodul von Erd- und Steinschüttdämmen. *Bautechnik* **69**, 515 – 517.
122. Vrymoed, J. (1981). Dynamic FEM model of Oroville Dam. *Proceedings of the ASCE* **107**, GT 8, 1057 – 1077.
123. Scott, R. F. (1973). The calculation of horizontal accelerations from seismoscope records. *Bulletin of the Seismological Society of America* **63**, 1637 – 1661.
124. Herzog, M. (1985). Das Verhalten des 235 m hohen Oroville-Staudamms bei den Erdbeden im Sommer 1975. *Wasserwirtschaft* **75**, 318 – 321.
125. Meidal, P. and Webster, J. L. (1973). Mica — one of the world's largest structures. *Water Power & Dam Construction* **25**, 201 – 210, 245 – 249.
126. Skermer, N. A. (1975). Mica Dam embankment stress analysis. *Proceedings of the ASCE* **101**, GT 3, 229 – 242.
127. Anonymous (1949). *Report on Investigations of failure of Sheffield Dam, Santa Barbara.* US Army Corps of Engineers.
128. Byerly, P. (1925). Notes on the intensity of the Santa Barbara Earthquake between Santa Barbara and San Luis Obispo. *Bulletin of the Seismological Society of America* **15**, 279 – 301.
129. Willis, B. (1925). A study of the Santa Barbara Earthquake of June 29, 1925. *Bulletin of the Seismological Society of America* **15**, 255 – 278.
130. Wegmann, E. (1888 – 1927). *The Design and Construction of Dams.* Several editions. Wiley & Sons, New York.
131. Marcello, C. (1956). Dighe a gravità alleggerite. *Wasser- & Energiewirtschaft* **48**, 206 – 212.

132. Herzog, M. (1993). Zutreffende Talsperrenbemessung mit einfachen Mitteln. *Österr. Ing. & Arch.-Zeitschrift* **138**, 16 – 26.
133. Herzog, M. (1990). Parameterstudie des Tragverhaltens einer grossen Gewölbestaumauer. *Wasserwirtschaft* **80**, 312 – 324.
134. Schweizerische Talsperrenkommission (1946). *Messungen, Beobachtungen und Versuche an schweizerischen Talsperren 1919 – 1945*. Eidg. Oberbauinspektorat, Bern.
135. Rescher, O. J. (1958). Talsperren im Kanton Wallis (Schweiz). *Österr. Ing.-Zeitschrift* **103**, 341 – 352.
136. Herzog, M. (1988). Der Einfluss der Blockfugen auf das Tragverhalten von Gewölbestaumauern. *Bauingenieur* **63**, 369 – 372.
137. Comité National Suisse des Grands Barrages (1964). *Comportement des Grands Barrages Suisses*. Bern, pp. 127 – 134.
138. Chopra, A. K. and Chakrabarti, P. (1972). The earthquake experience at Koyna Dam and stresses in concrete gravity dams. *Earthquake Engineering & Structural Dynamics* **1**, 151 – 164.
139. Davis, C. V. (1952). *Handbook of Applied Hydraulics*. McGraw-Hill, New York, 2nd edition pp. 23 – 46.
140. Raphael, J. M. (1952). The development of stresses in Shasta Dam. *Proceedings of the ASCE* **78**, No. 117, 1 – 20.
141. Grengg, H. (1962). Statistik 1961. *Die Talsperren Österreichs* No. 12. Österr. Wasserwirtschaftsverband, Vienna, pp. 121 – 126.
142. Steinböck, W. (1959). Die Staumauer am Grossen Mühldorfersee. *Die Talsperren Österreichs* No. 10. Österr. Wasserwirtschaftsverband, Vienna.
143. Mehmel, A. (1964). Verstärkungsmassnahmen an der Oleftalsperre. *Bauingenieur* **39**, 231 – 237.
144. Kelen, N. (1924). Einsturz der Gleno-Talsperre. *Beton & Eisen* **23**, 6 – 8, 133 – 138, 167 – 168.
145. Anonymous (1940). World's largest multiple arch dam completed. *Engineering News Record*, November 7, p. 39.
146. Mörsch, E. (1947). *Statik der Gewölbe und Rahmen*. Wittwer, Stuttgart.
147. Herzog, M. (1989). Die Traglast beliebiger Bögen in einfacher Näherung. *Österr. Ing. & Arch.-Zeitschrift* **134**, 467 – 474, 529.
148. Herzog, M. (1991). Entwicklung und gegenwärtiger Stand des Festigkeitsnachweises für Bogenstaumauern. *Österr. Ing. & Arch.-Zeitschrift* **136**, 126 – 132.
149. Tournadre, M. (1872). Description d'un barrage exécuté sur la rivière du Verdon et la prise d'eau du canal du Verdon en Provence. *Annales Ponts Chaussées* **42**, 1er sem., 428 – 455.
150. Herzog, M. (1989). Meilensteine des Baues von Bogenstaumauern. *Bautechnik* **66**, 73 – 80.
151. Wade, L. A. B. (1908 – 1909). Concrete and masonry dam construction in New South Wales. *Proceedings of the ICE* Part IV, **178**, 1 – 110.
152. Herzog, M. (1990). Einfache Formel für das Fusseinspannmoment von Gewölbestaumauern. *Wasser Energie Luft* **82**, 208 – 210.
153. Beyer, K. (1956). *Statik im Stahlbetonbau*. Springer, Berlin, 2nd edition, 2nd printing, pp. 783 – 784.
154. Coyne, A. (1937). The construction of large modern water dams. *Structural Engineer* **15**, 70 – 84, 331 – 338.
155. Semenza, C. (1956). Arch dams — development in Italy. *Proceedings of the ASCE* **82**, PO 3, Paper 1017, pp. 1 – 42.

156. Herzog, M. (1991). Die Rissgefahr im Innern von Bogenstaumauern. *Österr. Wasserwirtschaft* **43**, 69 – 70.
157. Herzog, M. (1992). Der Bruch der Bogenstaumauer Plum in China. *Bauingenieur* **67**, 405 – 407; Herzog, M. (1993). *Bauingenieur* **68**, 384.
158. The Engineering Foundation (1928). Report on arch dam investigation, Vol. I. *Proceedings of the ASCE*, Part 3.
159. Creager, W. P., Justin, J. D. and Hinds, J. (1945). *Engineering for Dams*. Wiley & Sons, New York, Vol. II, p. 81.
160. Herzog, M. (1982). Simplified analysis of concrete gravity and arch dams subjected to earthquakes. *Proceedings of the ICE*, Part 2, **73**, 189 – 196.
161. Nizery, A., Remenieras, G. and Beaujoint, N. (1963). Etude sur modèle réduit des contraintes dans les barrages. *Annales Ponts et Chaussées* Juillet – Août 1963, 395 – 437, 534 – 581.
162. Oberti, G. (1948). Diga del Lumiei. *L'Energia Elettrica* **25**, 409 – 431.
163. Herzog, M. (1979). Die statische Bruchfestigkeit von Gewölbestaumauern. *Bautechnik* **56**, 235 – 241.
164. Schnitter, G. (1948). Der Bau der Staumauer Rossens. *Schweiz. Bauzeitung* **66**, 641 – 644, 657 – 661, 687 – 691.
165. Tremmel, E. (1958). Limbergsperre — Statistische Auswertung der Pendelmessungen. *Die Talsperren Österreichs* No. 7. Österr. Wasserwirtschaftsverband. Vienna.
166. Herzog, M. (1988). Die jährliche Kronenbewegung von Gewölbestaumauern. *Bautechnik* **65**, 73 – 79.
167. Buchegger, W., Niederl, H., Petzny, H. and Widmann, R. (1964) Neuere Beobachtungen. *Die Talsperren Österreichs* No. 14. Österr. Wasserwirtschaftsverband, Vienna.
168. Rambert, O. and Verrey, A. (1962). Le barrage de Mauvoisin. *Bulletin Technique de la Suisse Romande* **88**, 369 – 374.
169. Elektro-Watt AG (1959). Erfahrungen beim Betrieb des Kraftwerks Mauvoisin. *Schweiz. Bauzeitung* **77**, 645 – 654.
170. Gilg, B., Sinniger, R., Gavard, M., Torrione, J. and Stucky, J. P. (1982). Long-term measurements on three Swiss dams — Mauvoisin, Grande Dixence and Mattmark. *14th ICOLD, Rio de Janeiro* Vol. II, pp. 909 – 926 (Q 52/R 54).
171. Herzog, M. (1991). Vollständige Näherungsberechnung von Gewölbestaumauern im Vergleich mit in situ-Messungen. *Bautechnik* **68**, 160 – 172.
172. Fenz, B. and Schenk, T. (1992). Die Erhöhung der Staumauer Mauvoisin. *Wasser Energie Luft* **84**, 245 – 248.
173. Obernhuber, P., Schöberl, P. and Wagner, E. (1994). Erhöhung der Sperre Limberg. *Österr. Ing. & Arch.-Zeitschrift* **139**, 526 – 534.
174. Mauboussin, G. (1955). The Gage dam. *Supplement 5th ICOLD, Paris.* Travaux No. 247, pp. 65 – 70.
175. Bellier, J. (1955). The project for a dam at Gage. *Supplement 5th ICOLD, Paris.* Travaux No. 247, pp. 62 – 64.
176. Frey, J. P. and Chauvet, P. (1957). La voûte très mince du barrage du Gage. *Annales ITBTP* **10**, 537 – 555.
177. Bellier, J., Mauboussin, G. and Mladyenovitch, V. (1958). Divers renseignements sur le comportement du barrage du Gage. *6th ICOLD, New York* Vol. II, pp. 1049 – 1062 (Q 21/R 103).
178 Dargeou, J. (1955). The Malpasset dam. *Supplement 5th ICOLD, Paris.* Travaux No. 247, pp. 151 – 154.

179. Mary, M. and Chardonnet, M. (1961). Zuschrift zu Talsperrenprobleme. *Österr. Wasserwirtschaft* **13**, 79 – 81.
180. Ru Naihua (1995). *Arch dams — Accidents and Safety of Large dams.* Shanghai, pp. 90 – 146 (in Chinese).
181. Gicot, H. (1961). Conceptions et techniques de quelques barrages voûtes suisses. *Wasser- & Energiewirtschaft* **53**, 194 – 205.
182. Herzog, M. (1981). Felssetzungen infolge Bergwasserabsenkung. *Bautechnik* **58**, 151 – 153.
183. Berchten, A. R. (1985). Repairs of Zeuzier arch dam in Switzerland. *15th ICOLD, Lausanne* Vol. II, pp. 693 – 711 (Q 57/R 40).
184. Gicot, H. and Lardy, P. (1955). Calculs et essais sur modéle réduit d'un barrage-voûte asymétrique. *5th ICOLD, Paris* Vol. III, pp. 1015 – 1038 (C 25).
185. Widmann, R., Stäuble, H., Klemen, K., Schlosser, J. and Gatti, H. (1979). Die Gewölbemauer Kölnbrein. *Österr. Zeitschrift f. Elektrizitätswirtschaft* **32**, 24 – 36.
186. Baustädter, K. and Widmann, R. (1985). The behaviour of the Kölnbrein arch dam. *15th ICOLD, Lausanne* Vol. I, pp. 633 – 651 (Q 57/R 37).
187. Lombardi, G. (1988). Querkraftbedingte Schäden in Bogensperren. *Wasser Energie Luft* **80**, 119 – 125.
188. Sargious, M. (1961). Hauptzugkräfte am Endauflager vorgespannter Betonbalken. *Bautechnik* **38**, 91 – 97.
189. Herzog, M. (1996). Hochgleiten von Bogenstaumauern auf flachen Talflanken. *Bautechnik* **73**, 829 – 831.